# ADVANCES IN NUCLEAR FUEL

Edited by **Shripad T. Revankar**

**Advances in Nuclear Fuel**
Edited by Shripad T. Revankar

**Published by InTech**

Janeza Trdine 9, 51000 Rijeka, Croatia



**Publishing Process Manager** Ivona Lovric

**Technical Editor** Teodora Smiljanic

**Cover Designer** InTech Design Team

Additional hard copies can be obtained from orders@intechopen.com

Advances in Nuclear Fuel, Edited by Shripad T. Revankar
p. cm.
ISBN 978-953-51-0042-3

For EU product safety concerns:
IN TECH d.o.o., Prolaz Marije Krucifikse Kozulić 3, 51000 Rijeka, Croatia,
info@intechopen.com or visit our website at intechopen.com.

# Meet the editor

Dr Revankar is a Professor of Nuclear Engineering and Director of Multiphase and Fuel Cell Research Laboratory at the School of Nuclear Engineering, Purdue University, West Lafayette, Indiana. He is also a WCU Visiting Professor in the Division of Advanced Nuclear Engineering at POSTECH, Republic of Korea. His current research interests are in nuclear fuel development, multiphase flow and heat transfer, microgravity multiphase flow in packed beds, fuel cell design, simulation and power systems, thermochemical water splitting hydrogen production, reactor thermalhydraulics and safety. Dr Revankar has published over 250 peer reviewed papers in archival journals and conference proceedings and is an active member of the American Nuclear Society, American Society of Mechanical Engineers (ASME), American Institute of Chemical Engineers (AIChE), Korean Nuclear Society (KNS) and Indian Society of Heat and Mass Transfer (ISHMT). He is a Fellow of ASME and has hosted various ASME and ANS conferences as session organizer, topical leader and track leader.

# Contents

# Preface

The socio economic growth rate of a nation is generally measured in terms of the per capita consumption of electric energy. An uninterrupted and sufficient power generation is required for the overall development and progress of any nation. Currently large scale electric energy production is primarily from fossil fuel (coal, oil, and natural gas), hydro and nuclear reactors. Problems with oil peaking, pollution and greenhouse gas emission persist with continued use of fossil fuel. At least for the next few decades, there are only four realistic options for reducing carbon dioxide emissions from electricity generation: increased efficiency in energy utilization, expanded use of renewable such as wind and solar, reducing carbon dioxide emissions at fossil fueled power plants by switching from coal to natural gas or by transitioning to capture and permanent sequestration of the carbon dioxide, and nuclear power. Though renewable power sources (solar, photovoltaic, wind, tidal and biomass) are desirable but they make up very small portion of energy portfolio. The uncontrollable capacity fluctuations and high production cost have contributed to smaller production capacities of the renewable energy.

Nuclear energy in some form promises abundant, safe, environmentally friendly, and economic power. Nuclear power prices compete well with those of fossil fuel and are less liable to fluctuate. Nuclear power also has the shortest energetic amortizing time and the highest energy yield of the energy sources compared in this respect. However for nuclear power to be sustainable the nuclear fuel cycle must be sustainable and there should be adequate nuclear fuel waste management program. The increasing use of nuclear power puts stringent demands on nuclear fuel cycle and waster management activities and on understanding the behavior of materials used in nuclear reactors.

To enable continued expansion of nuclear power, one must overcome critical challenges in cost, waste disposal, and proliferation concerns while maintaining its currently excellent safety and reliability record. In the relatively near term, important decisions may be taken with far reaching long-term implications about the evolution of the nuclear fuel cycle and waste management. Some of these include development of new fuels that have better safety features, sustainability through various fuel cycles including thorium fuel, fuel material operational issues such as material corrosion and crud formation on cladding, fuel behaviors during reactor accident, recycling of spent fuel and actinide burning, waste fuel material decontamination during and after nuclear reactor accidents, fuel and radioactive material proliferation, and safety. The present book published by InTech addresses these very key issues through eight chapters:

1. Numerical Analysis of Melting/Solidification Phenomena Using a Moving Boundary Analysis Method X-FEM
2. Relativistic Density-Functional Study of Nuclear Fuels
3. Radiation Induced Corrosion of Nuclear Fuel and Materials
4. Long Term Sustainability of Nuclear Fuel Resources
5. Behaviors of Nuclear Fuel Cladding During RIA
6. First Principles Simulations on Surface Properties and Oxidation of Nitride Nuclear Fuels
7. Nuclear Accidents in Fukushima, Japan, and Exploration of Decontamination $^{137}$Cs-Contaminated Soils
8. Introduction of Criticality Accident Evaluation

This book does not aim at a complete coverage or a broad range of topics on nuclear fuels. Instead, the goal of the book is to bring key issues associated with nuclear fuel material, advances in technology and its application in nuclear plants. The chapters focus on challenges and issues that need to be addressed for better performance and safety of nuclear fuel in nuclear plants. These focused areas are on development of high conductivity new fuels, radiation induced corrosion, fuel behavior during abnormal events in reactor, and decontamination of radioactive material. Authors hope to share ideas and findings so that new ideas and directions can potentially be developed focusing on nuclear fuel development and its application to nuclear power plants.

Worldwide the nuclear fuel technology has robust development program with are more than 430 nuclear power plants operating in the world and more plants are being constructed or planned for construction. Continuous technological advances will lead towards sustainable nuclear fuel through closed fuel cycles and advance fuel development. This book targets everyone - not only those interested in nuclear power but also for those who are concerned about the nuclear fuel waste management and radioactive material dispersal to environment. The book will hopefully inspire future R&D efforts, innovation by stimulating ideas on nuclear fuel material development, application and waste management.

We hope our readers will enjoy the book and will find it, both, interesting and useful.

**Dr. Shripad T. Revankar**
School of Nuclear Engineering, Purdue University, West Lafayette,
USA
&
Division of Advanced Nuclear Engineering,
Pohang University of Science and Technology, Pohang,
South Korea

# 1

# Long Term Sustainability of Nuclear Fuel Resources

Dubravko Pevec, Vladimir Knapp and Krešimir Trontl
*University of Zagreb, Faculty of Electrical Engineering and Computing*
*Croatia*

## 1. Introduction

The basic issue in considering the contribution of nuclear power to solving the world's energy problem in the future is the availability of uranium resources and its adequacy in meeting the future needs of nuclear capacity. Increased interest in nuclear energy is evident, and a new look into nuclear fuel resources is relevant. Sufficiency of nuclear fuel for the long-term use and expansion of nuclear power has been discussed by individual analysts and by institutions, with wide spectrum of answers corresponding to variety of initial assumptions on uranium resources, reactor technologies and energy strategies (Fetter, 2009; Nifenecker, 2003; Pevec et al., 2008). With a suitable choice of assumptions arguments were occasionally constructed for the claim that nuclear power has no long-term future due to inadequate fuel resources. Oppositely, again with appropriate choice of assumptions, reassuringly long times of nuclear fuel availability were obtained, even with inefficient once-through open nuclear fuel cycle. Typical such scenarios assume extension of present slow growth of nuclear power or assume a constant share of nuclear power in the total world energy production, now slightly above 6%. With once-through fuel cycle resources then may last well over a hundred years, as will be shown below, and, argument goes on, by that time we will have nuclear fusion, so there is no reason for concern about nuclear fuel. At present state of world affairs we cannot afford the comfort of such reasoning as it neglects the outstanding potential of nuclear energy to contribute to the solution of the probably crucial problem facing humanity; how to stop climate changes threatening our civilisation, and how to do it urgently. Unlike various alternative $CO_2$ non-emitting energy sources, fission energy is technically developed and available now, as witnessed by close to 430 reactors in operation (Knapp et al., 2010).

The nuclear energy has some characteristics different from fossil fuel energy which are very important when considering the long term sustainability of nuclear fuel resources.

First, unlike in the case of fossil fuels, the amount of energy obtainable from the resources per unit mass of nuclear fuel is far from being a fixed figure. Energy content of a kg of the standard coal is 29. 3 MJ. It is usable with high percentage of this figure for heating and with 30-40% if converted to electricity. Energy that can be obtained from a kg of natural uranium is highly dependent on the reactor type and on the nuclear fuel cycle. Presently dominant are so called thermal reactors. Their physically most important feature is that they fission practically only uranium isotope U235 which is present in natural uranium in only 0.7%. By presence of moderator in the core of these reactors fission neutrons are thermalised and thereby fission probability of U235 is increased by a large factor. Due to their even-even proton-neutron structure U238 nuclei can be fissioned only by fast neutrons. However, they can absorb slow neutrons and through two decays then U239 transform into a fissionable isotope of plutonium, Pu239, with properties similar to those of U235. As U235, it is fissionable by thermal neutrons and energy release per fission only slightly higher, some 2%. Consequently, in thermal reactors by neutron absorption a small amount of U238 is converted in plutonium, mainly Pu239. Plutonium is partly burnt in parallel with U235 and partly remaining in spent fuel. The thermal energy obtained per unit mass of the fuel in present thermal reactors is given in Table 1. Much the largest part of dominant isotope U238 remains unused. If plutonium is extracted from the spent fuel it can be added to the fresh fuel thereby increasing the amount of energy obtained from the unit weight of natural uranium. As the content of U238 in uranium is 99.3%, clearly a dramatically larger quantity of energy would be obtained if the energy of this isotope could be released (Bodansky, 2004).

| Fuel | Enrichment | Energy per Unit mass | | |
|---|---|---|---|---|
| Nat. uranium | 0.711% | 584 GJ/kg | 6.8 MWd/kg * | 7.3 MWd/kg ** |
| Enriched U | 3.5% | 2870 GJ/kg | 33.3 MWd/kg * | 36-40 MWd/kg ** |
| Plutonium | 100% | 82100 GJ/kg | 950 MWd/kg | |

* from fission of U235 only ** in reactor, with contribution from plutonium

Table 1. The thermal energy obtained per unit mass in present thermal reactors

Second, contribution of uranium cost to the cost of nuclear-generated electricity is low (2-4%) compared to contribution of fossil fuel cost to the cost of electricity of fossil power plant (25% for coal and 65% for gas). It follows that, even for conservative approach of 4% uranium cost contribution to electricity cost, five-fold increase in uranium cost would increase the cost of electricity by 16%, and ten-fold increase in uranium cost would have modest effect on the cost of electricity by increase of 36%. It will be shown that these large increases in uranium price would produce much larger increases in available uranium resources. These uranium resources will be sufficient to support an inefficient once-through fuel cycle till the end of the century and beyond, even in the case of rapid nuclear capacity growth.

Third, the operational lifetime of nuclear power plants is considerably longer than fossil power plant operational lifetime. The operational lifetime of current nuclear power plants is 40-60 years, and for Generation III nuclear power plants it will be 60-80 years. Therefore, the changes in nuclear fuel utilization will slowly change for long time periods.

In this chapter we address the issue of nuclear fuel resources long term sustainability in relation to the expected and projected high limit of growth of the world nuclear power. Three main aspects have to be analyzed in order to estimate how long the world's nuclear fuel supplies will last: nuclear fuel resources (uranium and thorium), technologies for nuclear fuel utilization, and energy requirements growth scenarios including different scenarios for nuclear share growth.

In the second section of this chapter conventional and unconventional uranium and thorium resources were presented and discussed. Figures given are valid for particular moment of time, with the rate of change of estimates dependent on the intensity of research and exploration.

Detailed analysis of potential technologies for improved nuclear fuel utilization is required in order to assess long term sustainability of nuclear fuel resources. Nowadays, thermal converter reactor technology with once-through nuclear fuel cycle is dominant. The effectiveness of the technology can be improved in the area of enrichment process as well as by introducing reprocessing of the spent fuel on larger scale. Other technologies are also on the development stage that allows their implementation in short or medium period of time. These include: thermal and fast breeder reactors of different kind, thorium based fuel cycle, and conversion of uranium or thorium by particle accelerators or fusion devices. The potential technologies for improved nuclear fuel utilization are analyzed in the third section.

Very important aspect of long term sustainability of nuclear fuel resources are scenarios for energy requirements growth, and scenarios for growth of nuclear share in electricity production resulting in overall nuclear capacity growth. The low growth scenario, the high growth scenarios with exponential and linear increases, and the scenario based on a compromise between low and high growth assumptions are presented in the fourth section.

The long term sustainability of nuclear fuel resources is discussed in fifth section, and the conclusions are given in the sixth section.

## 2. Nuclear fuel resources

Uranium, as well as thorium, can be used as a nuclear fuel.

Uranium is relatively abundant element in the upper earth's crust with the average content of 3 ppm. Uranium is a significant constituent of about hundred different minerals, but most minable ores belong to a dozen minerals (e.g. uraninite, davidite, uranothorite, carnotite, torbernite, autunite, etc.). Usually, uranium deposits are classified into four types: vein-type deposits, uranium in sandstones, uranium in conglomerates, and other deposits (pegmatites, phosphates).

The existing nuclear power reactors use uranium as a fuel. Uranium is natural element composed mainly of two isotopes U238 (99.27%) and U235 (0.72%). As the existing nuclear power reactors are thermal reactors, the bulk of the produced energy is obtained by fission of U235 isotope.

Thorium is three times more abundant element than uranium in the upper earth's crust with the average content of 6 - 10 ppm. Thorium is widely distributed in rocks and minerals, usually associated with uranium, elements of the rare-earth group and niobium and tantalum in oxides, silicates and phosphates. Thorium is natural element composed of only one isotope Th232 (100%). Although the Th232 isotope is not fissile, it can be converted to fissile isotope U233 by slow neutron absorption.

### 2.1 Uranium resources

Uranium resources are broadly classified as either conventional or unconventional. Conventional resources are those that have an established history of production where uranium is a primary product, co-product or an important by-product. Very low grade resources or those from which uranium is only recoverable as a minor by-product are considered unconventional resources.

Resource estimates are divided into separate categories according to different confidence level of occurrence, as well as on the cost of production.

### 2.1.1 Conventional uranium resources

The Red Book published in 2010 (Organization for Economic Co-operation and Development Nuclear Energy Agency [OECD/NEA] & International Atomic Energy Agency [IAEA], 2010) categorizes conventional uranium resources into Identified resources (corresponding to previously "Known conventional resources") and Undiscovered resources. Identified resources consist of reasonably assured resources and inferred resources. Reasonably Assured Resources (RAR) refers to uranium that occurs in known mineral deposits of delineated size, grade, and configuration such that the quantities which could be recovered within the given production cost ranges, with currently proven mining and processing technology can be specified. RAR have a high assurance of existence and they are expressed in terms of quantities of uranium recoverable from minable ore.

Inferred Resources (IR) refers to uranium, in addition to RAR, that is inferred to occur based on direct geological evidence, in extension of well-explored deposits, or in deposits in which geological continuity has been established but where specific data, including measurements of the deposits, and knowledge of the deposit's characteristics, are considered to be inadequate to classify the resource as RAR. The estimates in this category are less reliable than those in RAR. IR is corresponding to Estimated Additional Resources Category I (EAR-I) used up to the year 2003. IR is expressed in terms of quantities of uranium recoverable from minable ore.

Undiscovered resources include Prognosticated resources and Speculative resources.

Prognosticated Resources (PR) refers to uranium, in addition to Inferred Resources, that is expected to occur in deposits for which the evidence is mainly indirect and which are believed to exist in well-defined geological trends or areas of mineralisation with known deposits. Estimates of tonnage, grade and cost of discovery, delineation and recovery are based primarily on the knowledge of deposit characteristics in known deposits within the respective trends of areas and on such sampling, geological, geophysical or geochemical evidence as may be available. The estimates in this category are less reliable than those in IR. PR is corresponding to Estimated Additional Resources Category II (EAR-II) used up to the year 2003. PR is expressed in terms of uranium contained in minable ore, i.e., in situ quantities.

Speculative Resources (SR) refers to uranium, in addition to Prognosticated Resources, that is thought to exist, mostly on the basis of indirect evidence and geological extrapolations, in deposits discoverable with existing exploration techniques. The location of deposits envisaged in this category could generally be specified only as being somewhere within a given region or geological trend. Existence and size of such resources are speculative. SR is expressed in terms of uranium contained in minable ore, i.e., in situ quantities.

The Identified resources amount to 6.306 million tonnes (4.004 million tonnes of RAR and 2.302 million tonnes of Inferred resources). The Undiscovered resources amount to 10.401

million tonnes (2.905 million tonnes of Prognosticated resources and 7.496 million tonnes of Speculative resources). These estimates refer to uranium recoverable at cost of less than 260 USD/kg. Total conventional resources amount to 16.707 million tonnes according to Red Book as of January 2009. The Identified conventional resources for different cost ranges are given in Table 2. The Undiscovered conventional resources for different cost ranges are given in Table 3.

| **Resource category** | **Cost ranges** | | | |
|---|---|---|---|---|
| | **< 40 USD/kgU** | **< 80 USD/kgU** | **< 130 USD/kgU** | **< 260 USD/kgU** |
| Identified Resources (Total) | 796 | 3742 | 5404 | 6306 |
| Reasonably Assured Resources (RAR) | 570 | 2516 | 3525 | 4004 |
| Inferred Resources (IR) | 226 | 1226 | 1873 | 2302 |

Table 2. Identified conventional resources for different cost ranges in the year 2009 (1000 tU)

| **Resource category** | **Cost ranges** | |
|---|---|---|
| | **< 130 USD/kgU** | **< 260USD/kgU** |
| Undiscovered Resources (Total) | 6553 | 10401 |
| Prognosticated Resources (PR) | 2815 | 2905 |
| Speculative Resources (SR) | 3738 | 7496 |

Table 3. Undiscovered conventional resources for different cost ranges in year 2009 (1000 tU)

Countries with major uranium resources are Australia, Kazakhstan, Russian Federation, Canada, Niger, South Africa, USA, Namibia, and Brazil.

### 2.1.2 Unconventional uranium resources

Unconventional uranium resources (Barthel, 2007) are found in low grade deposits, or are recoverable as a by-product. Low grade uranium deposits in black shales, lignites, carbonatites and granites were expected to be potential sources in the past. However, developing a cost effective, environmentally acceptable means of uranium extraction from this potential source remains a challenge. By-product resources are of interest in the case that conventional resources are insufficient. In by-product recovery, the greatest portion of the costs is borne by the main products.

The most important unconventional uranium resources reported in Red Book 2010 (OECD/NEA & IAEA, 2010) are phosphate deposits and seawater.

#### 2.1.2.1 Phosphate deposits

At higher cost, uranium can be extracted from phosphate deposits. Uranium contained in phosphate deposits is estimated at 22 million tonnes, although annual production is limited by annual phosphoric acid production. The upper limit is below 10 000 t/year, so even if all the phosphoric acid production over time were considered, the total addition would not exceed one million tonnes. The historical operating costs for uranium recovery from phosphoric acid range from 60 to 140 USD/kgU (World Information Service of Energy [WISE], 2010). Recently, a new process (PhosEnergy) is being developed by Uranium Equities Limited, offering uranium recovery costs in the range from 65 to 80 USD/kgU. Design and construction of the demonstration plant is complete. It is expected to be in operation from late 2011 (World Nuclear Association [WNA], June 2011b). However, should uranium extraction, decoupled from phosphoric acid production, cost less than 200 USD/kgU an abundant addition to conventional resources would become available.

We do not assume that this will happen much sooner than 2060 and, thus, base our consideration on estimated conventional resources.

**2.1.2.2 Uranium from the seawater**

The uranium concentration in seawater is only 0.003 ppm, yet it can be extracted. It would require the processing of huge volumes of seawater (about 350 000 t water for 1 kg U) and use large amounts of absorber. The cost of extraction from seawater can be regarded as the upper limit of the cost of uranium. The quantity of uranium in the sea is about 4 billion tonnes, exceeding any possible needs for thousands of years.

Research on uranium recovery from seawater has been going in Germany, Italy, Japan, the United Kingdom and the United States of America in 1970s and 1980s, but is now only known to be continuing in Japan. Recent Japanese research showed that uranium extraction from the seawater is technically possible. It has been developed on a laboratory scale by using either resins or other specific adsorbent. An extraction cost as low as 250 USD/kg U has been estimated, which is more than twice as high as the present spot market price (Tamada et al., 2006). Although this price appears high, and certainly is, it could be acceptable for fast breeders with a closed fuel cycle.

## 2.2 Thorium resources

The principal sources of thorium are deposits of the placer type (concentrations of heavy minerals in coastal sands), from which monazite and other thorium bearing minerals are recovered. Thorium often occurs in minerals that are mined for another commodity and thorium being recovered as a by-product. Thorium is present in seawater with a concentration of only about 0.00005 ppm, due primarily to the insoluble nature of its only oxide, $ThO_2$. Thus the recovery of thorium from seawater is not a realistic option.

Estimates of thorium resources have been given in Red Books since 1965. Classification of thorium resources is similar to uranium, e.g. Reasonably Assured Resources (RAR) and Estimated Additional Resources (EAR). The EAR is separated into EAR-Category I and EAR-Category II according to different confidence level of occurrence. Identified resources consist of RAR and EAR-I. Prognosticated thorium resources are EAR-II. Thorium resources were also classified according to cost of recovery (OECD/NEA & IAEA, 2010).

The total world thorium resources, irrespective of economic availability, are at present estimated at about 6 million tonnes. The thorium resources recoverable at a cost lower than

80 USD/kg are estimated at 4.5 million tonnes. The identified thorium resources amount to 2 million tonnes and the prognosticated thorium resources amount to 2.5 million tonnes.

Countries with major thorium resources are Commonwealth of Independent States (former Soviet Union countries), Brazil, Turkey, USA, Australia, and India.

Due to the fact that thorium is roughly three times more abundant than uranium in the earth's crust and that exploration of thorium resources is poor, it is to be expected that ultimately recoverable thorium resources will be much higher than uranium resources.

### 2.3 Long term perspectives of nuclear fuel resources

The nuclear fuel resources given in preceding sections are the today's resource estimates published in the Red Book, compendium of data on uranium and thorium resources from around the world (OECD/NEA & IAEA, 2010). It is interesting to compare resource estimates over time (OECD/NEA, 2006). The evolution of Identified Resources, RAR, and EAR-I/IR over time (1973 - 2009) recoverable at cost of less than 130 USD/kg is shown in Fig. 1.

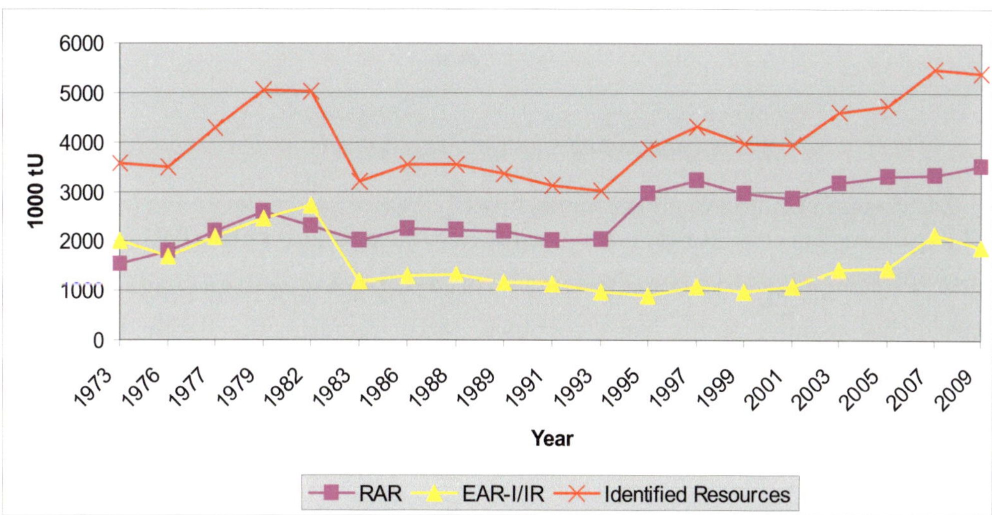


Fig. 1. Changes in Identified Resources, RAR, and EAR-I/IR over time (1973 - 2009)

The Identified Resources (including its components RAR and EAR-I/IR) mainly increased during a given time period except for a drop in year 1983. This drop could be explained by the facts that in year 1983 EAR have been subdivided into Category I and Category II and since 1983 RAR and EAR-I are given as recoverable resources(mining and milling losses deducted). The Identified Resources increased by around 60% in a time period of almost 40 years although for many years investment in exploration for uranium resources has been low.

The evolution of Undiscovered Resources, EAR-II/PR, Speculative Resources (< 130 USD/kgU), and Speculative Resources (regardless of the price) over time (1985 - 2009) is shown in Fig. 2.

The EAR-II/PR curve shows very gradual increase for the initial and final part of the given time period and for the rest of time period it remains nearly unchanged. That nearly unchanged part of the curve could be explained at least in part by the fact that countries tend to not re-evaluate their EAR-II/PR estimates on a regular bases. In contrast with the EAR-II/PR trends, both categories of Speculative Resources show considerably more volatility.

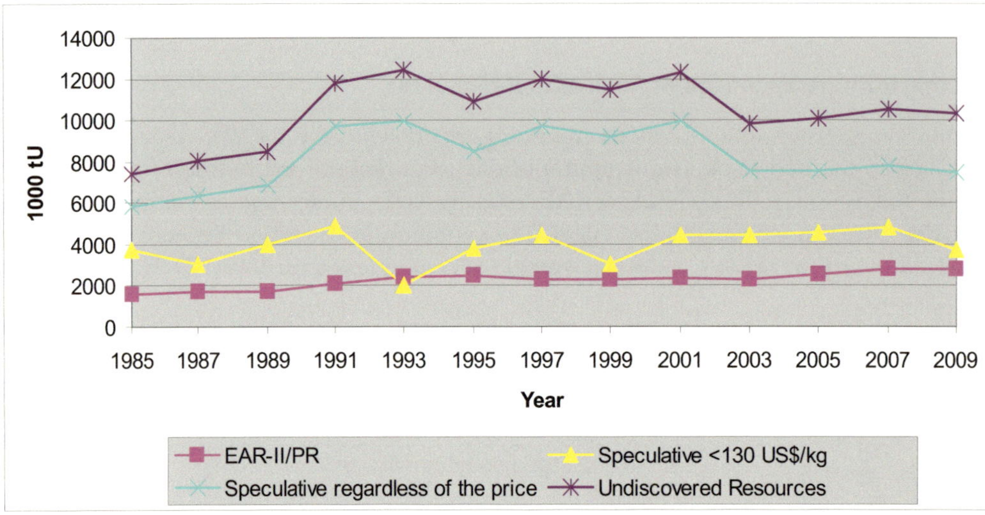


Fig. 2. Changes in Undiscovered Resources, EAR-II/PR, Speculative Resources (< 130 USD/kgU), and Speculative Resources (regardless of the price) over time (1985 - 2009)

The Red Book nuclear fuel resources estimates are obtained simply by collecting data on conventional resources from IAEA countries. Many countries are lightly explored in uranium and many countries do not report resources in all categories so there are almost certainly large quantities of uranium that are not yet included in Red Book. Therefore, the Red Book estimates of uranium resources should be considered a today's lower bound on the amount of uranium likely to be recoverable.

For analysis of uranium resources long term sustainability it is necessary to estimate the amount of uranium that will ultimately prove to be economically recoverable. This amount is defined as "total recoverable uranium resources". It depends on geologic parameters, as well as on development in technologies of exploration, extraction, and use. The total recoverable uranium resources could be determined from first principles by summarizing estimates of the abundance of uranium in the crust of the earth as a function of concentration and accessibility. Geologic data indicate that the total amount of uranium increases exponentially with decreasing ore grade. Synthesizing the power law for total amount of uranium and assumption that the cost of extracting a unit mass of uranium varies linearly with the inverse of the ore grade, one obtains a simple crustal model (Schneider & Sailor, 2008),

$$\frac{Q}{Q_0} = \left(\frac{P}{P_0}\right)^{\varepsilon}$$

where

$Q$ = quantity (MtU) of uranium available at the price level P (USD/kg U)
$Q_0$ = quantity of uranium available at some reference price $P_0$
$\varepsilon$ = long term elasticity of uranium supply.

This model must be calibrated through the selection of a reference point ($P_0$, $Q_0$) and the estimation of $\varepsilon$.

The Red Book data (OECD/NEA & IAEA, 2010) could be used as a reasonable point of departure for extrapolation of total recoverable uranium resources estimates. Therefore, the uranium resources quantity of 0,796 MtU available at price 40 USD/kgU (Table 2) has been selected as the reference point. The long term elasticity of uranium supply, $\varepsilon$, is estimated by different groups and its values range from 2.35 to 3.5. The World Nuclear Association (WNA, September 2011) concludes that a doubling in price from present levels could be expected to create about a tenfold increase in measured resources over time. It implies the long term elasticity of uranium supply, $\varepsilon$, to be equal 3.32. Another serious attempt to estimate how much uranium is likely to be available worldwide, based on Deffeyes and MacGregor (Deffeyes & MacGregor, 1980) distribution of uranium in the earth's crust, concluded that a ten-fold reduction in ore concentration is associated with a 300-fold increase in available resources. Using the assumption that costs are inversely proportional to ore grade the $\varepsilon$ value of 2.48 is obtained. The U.S. Department of Energy Generation IV Fuel Cycle Crosscut Group (FCCG) study (United States Department of Energy [USDOE], 2002), basing itself on the amounts of uranium recently estimated to be available in the United States at 30 USD/kgU and 50 USD/kgU, predicted that that the $\varepsilon$ might be as low as 2.35. Using the selected reference point and the obtained $\varepsilon$ values, total recoverable uranium resources are calculated by simple crustal model for different cost ranges. The calculated values and the Red Book values given in MtU are shown in Table 4. These values range from 4 MtU for cost category of 80 USD/kgU to almost 400 MtU for cost category of 260 USD/kgU. All of these estimates suggest that the total amount of uranium recoverable at prices 130 USD/kgU and 260 USD/kgU is likely to be substantially larger than the amount reported in the Red Book.

| **Source of estimate** | $\varepsilon$ | **Cost ranges** | | | |
|---|---|---|---|---|---|
| | | **< 40 USD/kgU** | **< 80 USD/kgU** | **< 130 USD/kgU** | **< 260 USD/kgU** |
| WNA | 3.32 | 0.796 | 7.96 | 39.84 | 397.91 |
| Deffeyes and MacGregor | 2.48 | 0.796 | 4.44 | 14.8 | 82.59 |
| Generation IV-FCCG | 2.35 | 0.796 | 4.06 | 12.70 | 64.75 |
| Red Book | | 0.796 | 3.742 | 5.402 | 6.306 |

Table 4. Total recoverable uranium resources estimated by simple crustal model for different cost ranges (MtU)

In our further analysis it was assumed that conventional uranium resources according to Red Book as of January 2009 in amount of 16.7 million tonnes will be recovered until year 2065.

Based on estimates obtained by simple crustal model we assumed that the total amount of uranium recoverable until the end of this century at still tolerable price of 180 USD/kgU is

50 Mt. This figure is supported by Update of the MIT 2003 Future of Nuclear Power study (Massachusetts Institute of Technology [MIT], 2009) in which "an order of magnitude larger resources are estimated at a tolerable doubling of prices".

## 3. Technologies for improvement of nuclear fuel utilization

A fuel utilization of present power reactors is low because they mainly utilize energy of U235 nuclide. Therefore, technologies and methods have been considered, that make possible to utilize enormous energy of U238 and of Th232 as well. Some of these technologies and methods are developed and proven technically viable, while some others are well researched with developmental problems identified. In the past, characterized by relatively slow nuclear energy expansion, with low cost of uranium and high cost of reprocessing of spent fuel, the simplest once-through fuel cycle has been generally accepted. Consequently, better utilization of nuclear fuel was not interesting to a private nuclear industry. Situation is different in the countries where governmental support made long term planning possible. For our purpose two aspects have to be understood. First, from the technical point of view, what these new technologies can achieve regarding uranium resources extension. Second important technical consideration is the time for their commercial development. It also has to be evaluated whether they could be introduced by the time of exhaustion of uranium resources used by present thermal reactors operating in the open cycle regime as is practice today. The following technologies and methods for improvement of nuclear fuel utilization have been considered:

a. Plutonium and uranium recycle with thermal reactor technology
b. Thermal breeder reactors
c. Fast breeder reactors
d. Zonal fuel burning in the so called „candle reactor"
e. Accelerator conversion of U238 into plutonium and of Th232 into U233
f. Conversion of U238 and Th232 by fusion neutrons

The short survey of each of the considered technologies and methods is given below.

### 3.1 Plutonium and uranium recycle with thermal reactor technology

Technology of plutonium recycle has been developing for many years. The PUREX process for recycling uranium and plutonium from spent nuclear fuel is implemented in several countries. Plutonium is mixed with enriched uranium for fabrication of the so called MOX fuel as both components are in the chemical form of oxides. There are many years of experience with the use of MOX fuel. Plutonium recycle is also a way to use surplus military plutonium. Except for such special situation, in the past there was little general interest in recycling at current high reprocessing and low uranium prices. Recent quantitative cost assessment of plutonium recycle has been given in EPRI Report 1018575 in 2009 (Electric Power Research Institute [EPRI], 2009). According to EPRI analysis fuel costs for once-through fuel cycle would be lower than for plutonium recycle for uranium cost below USD 312/kg and PUREX reprocessing cost above USD 750/kgHM. The same holds for uranium recycle except for some special concepts of reactors operating in tandem. The effect of plutonium and uranium recycle in present light water reactors on resources extension would not be very high; typically 5 kg of spent fuel contains enough plutonium for one kg of fresh fuel with plutonium replacing U235. Natural uranium resources extension is of the order of around 30%, as can be seen in a number of publications and reports (Garwin, 1998).

### 3.2 Thermal breeder reactors

Thermal breeder reactors were investigated in the early days of nuclear technology development before selection of light or heavy water cooled thermal reactors for commercial energy production. Thermal breeding is achieved either by benefiting from larger neutron yield of U233 in thermal fission, or by better neutron economy achieved by extracting neutrons absorbing fission products from the liquid fuel. First approach was investigated in the experimental Shippingport reactor. This light water solid fuel thermal breeder prototype reactor was in operation in US from 1957-1982 using uranium and thorium fuel, but the same concept could run on thorium fuel and U233 as fissile material produced by conversion of thorium. (United States Nuclear Regulatory Commission [USNRC], 2011). Other approach was also investigated in the early years of nuclear development. Small experimental molten uranium fluoride fuelled reactor (8 MW thermal power) was operated in the years 1965-69 at Oak Ridge National Laboratory in the US (Briggs, 1967; Rosenthal et.al., 1970). Using on-line extraction of fission products from circulating molten fuel, neutron losses by absorption in fuel were reduced with the effect of increasing conversion ratio above 1. Development did not proceed at the time due to corrosion problems. Latest development of this reactor type was Japanese FUJI MSR 100-200 MWe reactor. With several attractive features, such as reduced radioactivity inventory, low pressure of primary circuit, high thermal efficiency, possibility to run on thorium fuel, this concept is again taken up in a selection for Generation IV reactors. Corrosion problems were largely resolved in the meantime. Work on the molten salts technology is in progress in EU, China, India, with long interest in thorium, and other countries (Forsberg et.al., 2007; Gen. IV International Forum, 2011b).

Another concept of thermal breeder is a version of Canadian heavy water reactor CANDU using U233 as fissile material and thorium as fertile material. Commercial use of this fuel cycle, usable with little additional technical development required, depends on the costs of uranium and reprocessing of thorium for extraction of U233, and is ruled out at present uranium and reprocessing costs.

### 3.3 Fast breeder reactors

Concept of fast breeder reactor developed in early days of nuclear energy uses the physical property of Pu239 which when fissioned by fast neutrons releases considerably more fission neutrons than U235 or U233 fissioned at low or high neutron energy. Thus in reactor with Pu239 as fissile material and U238 as fertile, and with little or no moderation to avoid degradation of high neutron energy, conversion coefficient will be increased. With additional plutonium production by neutrons escaping from the reactor core into the uranium blanket surrounding the core, conversion ratio can reach values well above 1. Since these early days several concepts of fast reactors were developed to utilize energy of U238. One concept, sodium cooled fast reactors has been developed from the first small experimental reactor EBR 1 in USA, in operation 1951, to large reactors close to commercial stage, such as Superfenix of 1200 MW in France operating from 1984 to 1998, with a number of working prototypes in between in several countries. Last construction was reactor Monju of 300 MW in Japan, in operation from 1993. List of major experimental, pilot and demonstration fast breeder reactors is given in Table 5 (Cochran et.al., 2010; WNA, August 2011).

| Country | Name | MWe | MWth | Operation |
|---|---|---|---|---|
| China | CEFR | 20 | | 2010- |
| France | Rapsodie | | 40 | 1967-1983 |
| | Phenix | 250 | | 1973-2009 |
| | Superphenix | 1240 | | 1985-1998 |
| Germany | KNK 2 | 21 | | 1977-1991 |
| India | FBTR | | 40 | 1985- |
| | PFBR | 500 | | 2010? |
| Japan | Joyo | | 140 | 1977- |
| | Monju | 280 | | 1994-1995, 2010? |
| USSR/Russia | BR-5 | | 5 | 1959-2004 |
| | BOR-60 | 12 | | 1969- |
| | BN-350 (Kazakhstan) | 350 | | 1972-1999 |
| | BN-600 | 600 | | 1980- |
| | BN-800 | 800 | | 2014? |
| United Kingdom | Dounreay FR | 15 | | 1959-1977 |
| | Protoype FR | 250 | | 1974-1994 |
| United States | EBR-I | 0.2 | | 1951-1963 |
| | EBR-II | 20 | | 1963-1994 |
| | Fermi 1 | 66 | | 1963-1972 |
| | SEFOR | | 20 | 1969-1972 |
| | Fast Flux Test Facility | | 400 | 1980-1993 |

Table 5. Major experimental, pilot and demonstration fast breeder reactors

Other concepts of fast reactors using lead or lead-bismuth alloys as coolant, thus avoiding safety risks associated with sodium coolant, are selected as promising new projects for Generation IV reactors (Gen. IV International Forum, 2011a). Theoretical resource extension by fast breeder technology is very large, as the energy of dominant isotope of uranium is liberated. Extension is not only by a factor of about 50 coming from conversion of U238, but also from the possibility to use uranium resources too expensive for the present light water reactors with their inefficient use of uranium. It is correct to state that fast breeder reactors present technical option which can remove the resources constraint on any conceivable future nuclear energy strategy. Their deployment depends on economic and safety considerations, such as investment and reprocessing costs and plutonium diversion safety. New concepts in development attempt to preserve attractive safety features, such as low primary pressure, but avoid the use of sodium coolant which burns in contact with water.

### 3.4 Zonal fuel burning in so called "candle reactor"

Zonal burning concept, respectively, Travelling Wave Reactor (also called "candle reactor") (Ellis et.al., 2010) is an old idea proposed in 1958 by S. Feinberg (Feinberg, 1958). Recently it was given new attention by several investigators, especially by H. Sekimoto from the Tokyo

Institute of Technology (Sekimoto et.al., 2008). This reactor concept promises very high uranium utilization, about 40% of U238 in fuel, without the need for reprocessing. Needles to say, that would dramatically increase energy obtainable from uranium with very great advantage that reprocessing is not required. Fissile material is burnt and created in situ in the zone that moves through the reactor core. Concept is certainly very attractive, but real perspective is not yet clear. It could be a major advance in the use of nuclear fission energy, but it has not been demonstrated and is still in the early phase of development. Open problems are fuel and other core materials capable to sustain very high burn-up. Clarifications on the initiation of the burning are needed. Attempt to construct a prototype of this reactor type is supported by Bill Gates foundation.

### 3.5 Accelerator conversion of U238 into plutonium and of Th232 into U233

Electronuclear breeding investigation started early within the US MTA project (1949-1954), initiated by E.Lawrence (Heckrotte, 1977). Although technically successful, project was terminated when new uranium deposits large enough for US nuclear programme were discovered. Number of studies in 70ties dealt with the accelerator production of fissile materials Pu239 or U233, but low cost of uranium and proliferation consideration worked against further development. Concept was recently again taken up by C. Rubbia of CERN. In electronuclear accelerator breeding, particle accelerator is optimized in particle energy and target selection to produce thermal neutrons at minimum energy cost. Using protons in the range of 1000- 1500 MeV or deuterons with twice this value, minimum energy is lost on ionization in the large uranium or thorium target, whilst energetic ions produce neutrons first in spallation reactions and then in fast neutron reactions such as (n,2n) or (n,3n) which further increase number of neutrons of lower energy before they are thermalized and absorbed in fertile materials U238 or Th232. Project studies show that economy of plutonium production requires the proton beams of 200 300 mA corresponding to a beam power of about 200-300 MW. It is believed that extrapolation of present accelerators to such beams would not require new physical development. Accelerator target would in size and power dissipation resemble nuclear reactor core, profiting thereby from the existing reactor technology. Such an accelerator combined with the conventional thermal reactors fed by fertile nuclides produced by accelerator-breeder would present a system producing energy with an input of natural uranium or thorium fuel only. While in principle such hybrid system offers as effective use of natural uranium as a fast breeder reactor, it has an important advantage that fissile material production can be separated in time and location from the energy production. Accelerator and reprocessing installation would parallel enrichment installations, with a difference that the largest part of natural uranium input could be turned into fissile isotopes. Another advantage is that produced fissile materials could be fed into existing proven conventional reactors (Bowman et.al., 1992; Fraser et.al., 1981; Kouts & Steinberg, 1977; Lewis, 1969; Steiberg et al., 1983).

### 3.6 Conversion of U238 and Th232 by fusion neutrons

Several studies have shown that fusion devices unable to reach positive energy balance required to operate as pure fusion power producer, could still serve as neutron source producing neutrons for conversion of uranium or thorium. With fissile materials produced by neutron irradiation fed into conventional fission reactors, hybrid system of fusion device and fission reactors can produce energy with input of natural uranium or thorium only, as accelerator breeder systems. Many general and economic considerations are similar to those for accelerator breeders, with an advantage of less complexity in case of accelerator system, where the accelerator target technology could use much of reactor core technology. At present development required for accelerator breeders appears less demanding than development of fusion breeder devices (Maniscalco et.al., 1981).

### 3.7 Perspectives of nuclear fuel utilization improvement

At this moment it is difficult to foresee which, if any, of these ways to utilize the energy of U238 and Th232 will be developed. Molten salt thermal breeder might have the best chance, being one of the Generation IV selections. Second chance could be one of the fast breeder concepts with the coolant more acceptable than sodium. When we look at the technologies that may require more time for development, such as accelerator breeders or fusion –fission hybrids, we should note that time is not a limitation, as with effective burning of U238 nuclear fission energy is a source for the next thousands of years. At that time scale it does not matter whether they are developed in 50 or in 100 years. What is however important is to know that technologies exist which if developed and applied would make nuclear fission an energy source we cannot run out.

Cost of enriched sea extracted uranium determines the upper limit on the costs of any of above concepts for utilization of U238. An essential reduction of seawater uranium extraction cost would consequently reduce the number of economically acceptable concepts out of the list of physically and technically possible concepts presented above, respectively, move them into the more distant future.

## 4. Projections of long term world nuclear energy demand and nuclear fuel requirements

In order to assess long term sustainability of uranium resources a number of scenarios with different nuclear energy development strategies have been analysed. In the upcoming subsection we first give general assumptions and calculational methodology used in the analysis of all scenarios. We then proceed with detailed description of each particular scenario including specific assumptions and overall calculational results.

### 4.1 General assumptions and calculational methodology

In all the development strategies, i.e., scenarios, once-through fuel technology has been used. Spent fuel was assumed to be stored in spent fuel casks on controlled sites, enabling possibility of future reprocessing. The year 2010 has been chosen as the starting year for all the scenarios. The initial parameters used are those for the year 2009 and are based on the World Energy Outlook (WEO) 2009 (IEA, 2009) reference scenario data, the joint report by OECD Nuclear Energy Agency and the International Atomic Energy Agency regarding uranium resources (OECD/NEA & IAEA, 2010), and some assumptions based on engineering judgement and experience. These parameters are as follows:

- conventional uranium resources have been used in all scenarios as availability merit; these resources equal to 16.7 million tonnes (OECD/NEA & IAEA, 2010),

- conversion factor addressing the amount of uranium required for production of 1 TWh of electricity equals 25.0 tU/TWh; the factor has been conservatively set based on the analyses of electricity production in nuclear power plants and corresponding uranium demand over the last decade (OECD, 2006; OECD/NEA & IAEA, 2010); the value for the conversion factor has been verified theoretically (Bodansky, 2004),
- conversion factor addressing the mass of plutonium in spent fuel based on energy production is 0.17 tPu/GWye (Bodansky, 2004),
- constant capacity factor for nuclear power plants of 0.88 has been used for the entire investigated period in all scenarios,
- scenarios 2, 3 and 4 are selected in order to see the adequacy of uranium resources for essential contribution to carbon emission reduction, as required by WEO 2009 450 Strategy that would keep temperature increase below 2°C (IEA, 2009). Owing to general safety consideration we assume conventional reactor technology until the end of century and postponement of reprocessing until 2065, respectively 2100. This is also the reason for using conservative parameters for evaluation of uranium consumption. Scenario 1 is a low growth scenario which would not contribute essentially to carbon emission reduction.

### 4.2 Scenario 1 – Low growth scenario

A scenario of low nuclear capacity growth is a typical scenario showing that for a small share of nuclear energy in the total world production of energy, resources are not a limiting factor. This scenario assumes moderate growth strategy of 0.6% per year for the period 2011 - 2025, and 1.3% after the year 2025, following the 450 Policy Strategy of WEO 2009 (IEA, 2009). The scenario aims at preserving the share of nuclear energy in the total energy production. Although the present growth of total energy production and consumption is higher, we do not consider it appropriate for the longer periods in question. The investigated period is the entire 21$^{st}$ century, with special attention placed on the year 2065, which is later used as a milestone in scenario 2 and scenario 3. The results are given in Table 6.

Cumulative uranium requirements up to the year 2065 would be approximately 5.4 million tonnes, while for the entire 21$^{st}$ century cumulative requirements would reach 11.3 million tonnes. By the year 2100 installed nuclear capacity would reach 1080 GWe producing more than 8,000 TWh of electricity per year. It is also interesting to notice that cumulative mass of plutonium in spent fuel by the year 2100 would be slightly below 9,000 tonnes. If the same level of nuclear capacity increase would be used beyond the year 2100, the conventional uranium resources of 16.7 million tonnes would be exhausted by the year 2123.

### 4.3 Scenario 2 – Exponential high growth scenario

Exponential high growth scenario is determined by asking for the maximum nuclear build-up that can be reached by the year 2065, compatible with present estimate of uranium resources and their use with once-through nuclear technology, i.e. without reprocessing. Exponential growth with annual increase of 2.35% is used for the initial period 2011 - 2025.

The aim of the scenario analysis is to deduce the maximum growth, i.e., the maximum nuclear build-up that can be achieved throughout the period 2026 - 2065, with the assumption that at the end of the period the current uranium resources of 16.7 million tonnes would be exhausted. The year 2026 has been chosen as the starting year for rapid nuclear build-up based on the estimate of present status of nuclear industry and the time needed to prepare such a massive undertaking. The results are given in Table 7.

| Year | Nuclear capacity (GWe) | Annual electricity production [TWh] | Annual U requirements (ktU) | Cumulative U requirements (ktU) | Annual mass of Pu in spent fuel (tPu) | Cumulative mass of Pu in spent fuel (tPu) |
|---|---|---|---|---|---|---|
| 2010 | 375 | 2,890 | 72 | 72 | 56 | 56 |
| 2015 | 386 | 2,978 | 74 | 440 | 58 | 342 |
| 2020 | 398 | 3,068 | 77 | 819 | 60 | 636 |
| 2025 | 410 | 3,161 | 79 | 1,210 | 61 | 939 |
| 2030 | 437 | 3,372 | 84 | 1,620 | 65 | 1,258 |
| 2035 | 467 | 3,597 | 90 | 2,059 | 70 | 1,598 |
| 2040 | 498 | 3,837 | 96 | 2,526 | 74 | 1,961 |
| 2045 | 531 | 4,093 | 102 | 3,025 | 79 | 2,348 |
| 2050 | 566 | 4,366 | 109 | 3,557 | 85 | 2,761 |
| 2055 | 604 | 4,658 | 116 | 4,124 | 90 | 3,201 |
| 2060 | 644 | 4,968 | 124 | 4,729 | 96 | 3,671 |
| 2065 | 687 | 5,300 | 132 | 5,375 | 103 | 4,172 |
| 2070 | 733 | 5,653 | 141 | 6,064 | 110 | 4,707 |
| 2080 | 834 | 6,433 | 161 | 7,582 | 125 | 5,886 |
| 2090 | 950 | 7,320 | 183 | 9,310 | 142 | 7,227 |
| 2100 | 1,080 | 8,329 | 208 | 11,276 | 162 | 8,753 |

Table 6. Scenario 1 (low growth scenario) results

Under the condition of uranium resources exhaustion by the year 2065, the maximum possible annual growth rate for the period 2025 – 2065 is 5.7%. Thus, by the year 2065 installed nuclear capacity would reach 4,878 GWe producing more than 37,000 TWh of electricity in that year. Under the scenario terms, the maximum increase of nuclear capacity is observed during the last year of examined period and equals 263 GWe. It is also interesting to notice that cumulative mass of plutonium in spent fuel until the year 2065 would slightly exceed 13,000 tonnes. Very high contribution, over 50%, to the carbon emission reduction as required by WEO 2009 450 Strategy would be reached by 2065.

Based on previous discussion on long-term perspective of nuclear fuel resources presented in subsection 2.3, one can assume that the current estimate of 16.7 million tonnes of conventional uranium resources is likely to increase in the next 50 years. Therefore, it would be interesting to see the uranium requirements for the entire 21st century. A number of development strategies for the period 2066-2100 could be taken into account. However, we limit our investigation on a simple one, foreseeing constant nuclear capacity that equals the one reached by the year 2065 - 4,878 GWe. The results are also given in Table 7.

Cumulative uranium requirements for the period 2066-2100 would amount to approximately 33 million tonnes. If reprocessing of spent fuel and plutonium cycle (MOX fuel) is envisioned as possible after the year 2065 (WNA, 2011a), then cumulative mass of plutonium in spent fuel up to the year 2098 would amount to slightly more than 37 thousand tonnes. The year 2098 has been taken as final for plutonium accumulation to enable reprocessing of spent fuel and MOX production. Assuming that 70% of accumulated plutonium in spent fuel is fissile (Bodansky, 2004) reduction of uranium requirements in the amount of 5.7 million tonnes could be expected.

| Year | Nuclear capacity (GWe) | Annual electricity production [TWh] | Annual U requirements (ktU) | Cumulative U requirements (ktU) | Annual mass of Pu in spent fuel (tPu) | Cumulative mass of Pu in spent fuel (tPu) |
|---|---|---|---|---|---|---|
| 2010 | 375 | 2,890 | 72 | 72 | 56 | 56 |
| 2015 | 421 | 3,246 | 81 | 460 | 63 | 357 |
| 2020 | 473 | 3,646 | 91 | 895 | 71 | 695 |
| 2025 | 531 | 4,095 | 102 | 1,384 | 79 | 1,074 |
| 2030 | 701 | 5,403 | 135 | 1,990 | 105 | 1,545 |
| 2035 | 925 | 7,128 | 178 | 2,790 | 138 | 2,166 |
| 2040 | 1,220 | 9,405 | 235 | 3,846 | 183 | 2,985 |
| 2045 | 1,610 | 12,409 | 310 | 5,238 | 241 | 4,066 |
| 2050 | 2,124 | 16,372 | 409 | 7,076 | 318 | 5,492 |
| 2055 | 2,802 | 21,601 | 540 | 9,500 | 419 | 7,374 |
| 2060 | 3,697 | 28,500 | 712 | 12,698 | 553 | 9,857 |
| 2065 | 4,878 | 37,603 | 940 | 16,918 | 730 | 13,133 |
| 2070 | 4,878 | 37,603 | 940 | 21,618 | 730 | 16,781 |
| 2075 | 4,878 | 37,603 | 940 | 26,319 | 730 | 20,430 |
| 2080 | 4,878 | 37,603 | 940 | 31,019 | 730 | 24,079 |
| 2085 | 4,878 | 37,603 | 940 | 35,719 | 730 | 27,727 |
| 2090 | 4,878 | 37,603 | 940 | 40,420 | 730 | 31,376 |
| 2095 | 4,878 | 37,603 | 940 | 45,120 | 730 | 35,025 |
| 2100 | 4,878 | 37,603 | 940 | 49,821 | 730 | 38,673 |

Table 7. Scenario 2 (exponential high growth scenario) results

### 4.4 Scenario 3 – Linear high growth scenario

As in the previous scenario, a scenario of linear high growth is determined by asking for the maximum nuclear build-up that can be reached by the year 2065 with the assumption that current conventional uranium resources would be exhausted by the same year. However opposed to scenario 2, it assumes linear growth rate. Also for the period 2011-2025 linear growth rate is envisioned similar to the WEO 2009 reference scenario (IEA, 2009) resulting in 459 GWe of installed nuclear capacity in the year 2025. Annual increase in nuclear capacity for the period 2011 - 2025 is approximately 5.6 GWe. The results of scenario 3 analysis are given in Table 8.

Under the same conditions as in the previous scenario (current uranium resources exhaustion by the year 2065), the maximum possible annual increase of installed nuclear capacity for the period 2025 – 2065 is 75.5 GWe. Thus, by the year 2065 installed nuclear capacity would reach 3,479 GWe producing almost 27,000 TWh of electricity per year. Compared to previous scenario, scenario 3 results in larger penetration of new nuclear capacity at the beginning of investigated period. This is an advantage from the carbon emission reduction considerations. For example, scenario 2 projects 30 GWe of new nuclear capacity for the year 2026, as opposed to 75.5 GWe of scenario 3. Graphical representation of annual increase in nuclear capacity for scenario 2 and scenario 3 is given in Fig. 3. Cumulative mass of plutonium in spent fuel until the year 2065 would slightly exceed 13,000 tonnes just as in the case of the previous scenario.

As well as for scenario 2, extension of scenario 3 up to the year 2100 has been analysed, assuming nuclear capacity of 3,479 GWe for the period 2066-2100. The results of extended scenario 3 are also given in Table 8.

| Year | Nuclear capacity (GWe) | Annual electricity production [TWh] | Annual U requirements (ktU) | Cumulative U requirements (ktU) | Annual mass of Pu in spent fuel (tPu) | Cumulative mass of Pu in spent fuel (tPu) |
|---|---|---|---|---|---|---|
| 2010 | 375 | 2,890 | 72 | 72 | 56 | 56 |
| 2015 | 403 | 3,106 | 78 | 450 | 60 | 350 |
| 2020 | 431 | 3,322 | 83 | 854 | 65 | 665 |
| 2025 | 459 | 3,538 | 88 | 1,286 | 69 | 1,001 |
| 2030 | 836 | 6,448 | 161 | 1,946 | 125 | 1,515 |
| 2035 | 1,214 | 9,358 | 234 | 2,971 | 182 | 2,312 |
| 2040 | 1,591 | 12,269 | 307 | 4,359 | 239 | 3,392 |
| 2045 | 1,969 | 15,179 | 379 | 6,110 | 295 | 4,756 |
| 2050 | 2,346 | 18,089 | 452 | 8,226 | 352 | 6,403 |
| 2055 | 2,724 | 20,999 | 525 | 10,705 | 409 | 8,332 |
| 2060 | 3,101 | 23,909 | 598 | 13,548 | 465 | 10,545 |
| 2065 | 3,479 | 26,819 | 670 | 16,755 | 522 | 13,041 |
| 2070 | 3,479 | 26,819 | 670 | 20,108 | 522 | 15,650 |
| 2075 | 3,479 | 26,819 | 670 | 23,460 | 522 | 18,260 |
| 2080 | 3,479 | 26,819 | 670 | 26,812 | 522 | 20,869 |
| 2085 | 3,479 | 26,819 | 670 | 30,165 | 522 | 23,478 |
| 2090 | 3,479 | 26,819 | 670 | 33,517 | 522 | 26,087 |
| 2095 | 3,479 | 26,819 | 670 | 36,869 | 522 | 28,697 |
| 2100 | 3,479 | 26,819 | 670 | 40,222 | 522 | 31,306 |

Table 8. Scenario 3 (linear high growth scenario) results

Cumulative uranium requirements for the period 2066-2100 would amount to approximately 23.5 million tonnes. The cumulative mass of plutonium in spent fuel up to the year 2098 would amount to 30.2 thousand tonnes. If reprocessing of spent fuel and plutonium cycle (MOX fuel) is envisioned as possible after the year 2065, and using the same assumption as in the previous scenario a reduction of uranium requirements in the amount of 4.6 million tonnes would be expected.

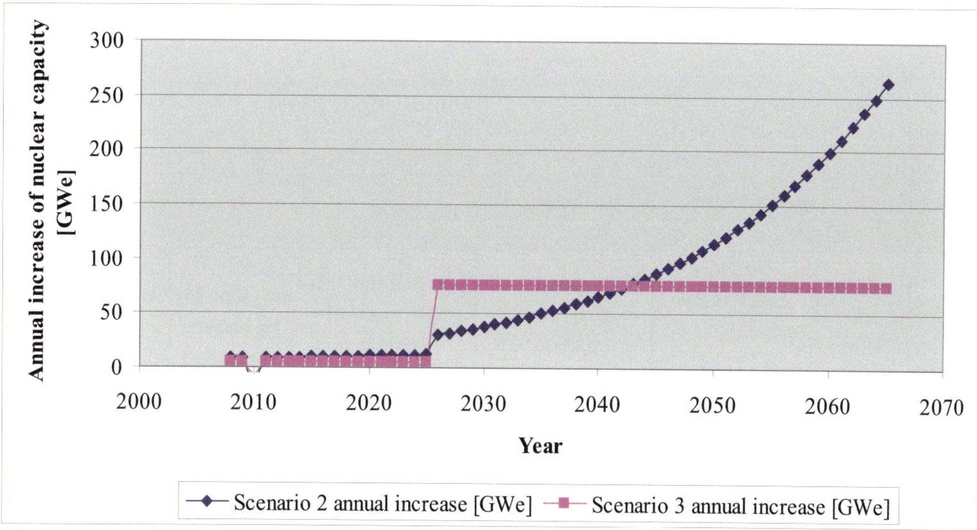


Fig. 3. Annual increase in nuclear capacity for scenario 2 and scenario 3

### 4.5 Scenario 4 – An intermediate scenario

Scenarios 2 and 3, i.e., high growth scenarios, provide illustration on maximum growth of nuclear capacities possible under stated resources constraint. Scenario 4 illustrates a less demanding nuclear build-up strategy that would replace all coal power plants without Carbon Capture and Storage (CCS) system, with nuclear power plants during the 2026-2065 period. Unlike scenario 1 this scenario would still give important contribution to carbon emission reduction, albeit not as high as the scenarios 2 and 3. It is assumed that all new coal power plants build after the year 2025 would have CCS installations. Linear replacement dynamics starting in the year 2026 is assumed without specifying the exact dates of coal power plant replacement. As in the previous scenario linear growth rate is envisioned for the period 2011-2025, similar to the WEO 2009 reference scenario (IEA, 2009), resulting in 459 GWe of installed nuclear capacity in the year 2025. Same WEO 2009 reference scenario (IEA, 2009) states that electricity production in coal power plants would be 13,387 TWh in the year 2025. With availability factor of 0.88, installed nuclear capacity of 1,736 GWe would be required to replace coal power plants electricity production. The results of scenario 4 analysis are given in Table 9.

Goal of all non-CCS coal power plants replacement throughout the period 2026-2065 would require an annual increase of nuclear capacity in the amount of 43.4 GWe. The total installed nuclear power by the year 2065 would reach 2,195 GWe with electricity production of almost 17,000 TWh. Cumulative mass of plutonium in spent fuel until the year 2065 would slightly exceed 9,000 tonnes which is rather lower than in previous two scenarios.

As in the previous two scenarios, extension of scenario 4 up to the year 2100 has been analysed assuming nuclear capacity of 2195 GWe for the period 2066-2100. The results of extended scenario 4 are also given in Table 9.

Intermediate nuclear growth envisioned in scenario 4 results in cumulative uranium requirements up to the year 2065 in the amount of slightly less than 12 million tonnes. The current conventional uranium resources would be exhausted by the year 2077. Cumulative uranium requirements for the period 2078-2100 would amount to approximately 9.7 million tonnes. If reprocessing of spent fuel and plutonium cycle (MOX fuel) is envisioned as possible after the year 2065, then cumulative mass of plutonium in spent fuel up to the year 2098 would amount to 19.9 thousand tonnes resulting in possible reduction of uranium requirements in the amount of 3.1 million tonnes.

| Year | Nuclear capacity (GWe) | Annual electricity production [TWh] | Annual U requirements (ktU) | Cumulative U requirements (ktU) | Annual mass of Pu in spent fuel (tPu) | Cumulative mass of Pu in spent fuel (tPu) |
|---|---|---|---|---|---|---|
| 2010 | 375 | 2,890 | 72 | 72 | 56 | 56 |
| 2015 | 403 | 3,106 | 78 | 450 | 60 | 349 |
| 2020 | 431 | 3,322 | 83 | 854 | 64 | 663 |
| 2025 | 459 | 3,538 | 88 | 1,286 | 69 | 998 |
| 2030 | 676 | 5,211 | 130 | 1,853 | 101 | 1,439 |
| 2035 | 893 | 6,884 | 172 | 2,630 | 134 | 2,042 |
| 2040 | 1,110 | 8,557 | 214 | 3,616 | 166 | 2,807 |
| 2045 | 1,327 | 10,230 | 256 | 4,811 | 199 | 3,735 |
| 2050 | 1,544 | 11,902 | 298 | 6,215 | 231 | 4,825 |
| 2055 | 1,761 | 13,575 | 339 | 7,829 | 263 | 6,077 |
| 2060 | 1,978 | 15,248 | 381 | 9,651 | 296 | 7,492 |
| 2065 | 2,195 | 16,921 | 423 | 11,683 | 328 | 9,069 |
| 2070 | 2,195 | 16,921 | 423 | 13,798 | 328 | 10,710 |
| 2075 | 2,195 | 16,921 | 423 | 15,913 | 328 | 12,352 |
| 2080 | 2,195 | 16,921 | 423 | 18,028 | 328 | 13,994 |
| 2085 | 2,195 | 16,921 | 423 | 20,143 | 328 | 15,636 |
| 2090 | 2,195 | 16,921 | 423 | 22,258 | 328 | 17,278 |
| 2095 | 2,195 | 16,921 | 423 | 24,373 | 328 | 18,920 |
| 2100 | 2,195 | 16,921 | 423 | 26,488 | 328 | 20,562 |

Table 9. Scenario 4 (intermediate growth scenario) results

## 5. Discussion on the long term sustainability of nuclear resources

As we stated introductory, energy that can be released by nuclear fission from uranium or thorium is not determined, or not essentially determined, by the quantity of resources. This is an essential difference to note when comparing nuclear with fossil fuel resources. On the other hand physical quantities of resources are, similarly as for fossil fuels, defined by extraction costs and by accepted criteria for categorization and estimates of deposits. Energy that can be liberated from unit mass of natural uranium varies by a large factor depending on the reactor and fuel cycle technology. Economic criteria on uranium deposits are consequently much more dependent on the energy conversion technology than in the fossil energy use. If the technology applied releases much more energy per unit mass than the present conventional reactors, then more expensive uranium or thorium deposits can be economically exploited. However our approach on the nuclear technologies to be used in this century is conservative. Therefore, our first interest is to see how far we can go with conventional, or essentially conventional nuclear technology. When considering present and future nuclear technologies which determine the requirements we must not take a narrow technical view on the possible fuel and reactor technologies. Development of nuclear safety is a slow process, reactors built in the nuclear boom in the late seventies and early eighties of the last century are still running, albeit approaching retirement. Although there are some 14 000 years of reactor experience, change of generations is a slow process, and such is the rate of change in basic reactor concepts. As the recent accidents at Fukushima show there is still a room for improvement even on the dominant line of light water reactors operating in a once-through fuel cycle. This is a reason why we estimate the uranium requirement in this century without introduction of breeder reactors. Also, we do not foresee before the end of century any major contribution of other technologies for extension of uranium or thorium utilization (Section 3). Our further basic assumption is on the role that nuclear fission should play in the critical period of about 50 years from now before wind, solar, nuclear fusion and CCS may contribute essential part of energy production. Nuclear fission energy is a proven, developed and economical source of carbon free energy. It is very difficult to see that the internationally accepted target to keep the mean global temperature increase below 2 °C could be achieved without the use of nuclear energy. Therefore in estimating the future needs of uranium we consider such deployments of nuclear power as can give an essential contribution to reduction of carbon emission. Often shown strategies with low growth, such as scenario 1 included in previous Section 4, result in assurances about the long life of resources, but are pointless for the purpose of climate control. For our purpose relevant are strategies 2, 3, and 4 of Section 4.

These strategies are an extension of the strategies we investigated earlier (Knapp et al., 2010) with the aim to determine what could be maximum contribution of nuclear energy in reduction of carbon emission down from the projected WEO 2009 Reference scenario to the sustainable WEO 450 scenario limiting the temperature increase to 2°C. Strategies were constrained to the use of proven conventional reactors operating in the once-through nuclear fuel cycle, without fuel reprocessing and plutonium recycle. Maximum nuclear contribution was obtained in strategies 2 and 3 by further assumption that total conventional uranium resources estimated in 2009 Red Book be consumed by the year 2065. The point of the study was not in proposing any specific growth strategy, but rather to see whether with conventional reactor technology, without spent fuel reprocessing, nuclear energy can essentially contribute to the carbon emission reduction. An argument for selection of the year 2065 for the final year of nuclear build-up is essentially derived from the status of nuclear and renewable technologies, as well as CCS and fusion prospects and their perspective for large contributions in carbon emission reduction. Under these constraints maximum annual nuclear capacity growth for the linear growth strategy (scenario 3), between the years 2025 and 2065 was 75.5 GW, reaching installed nuclear power of 3479 GW in 2065. By that year nuclear contribution to the required GreenHouse Gasses (GHG) emission reduction comes to the value of 39.6% of the WEO 450 Strategy requirements (Knapp et al., 2010). This is a very serious contribution which still leaves large space of remaining about 60%, respectively of 38.4 GtCO2-eq reduction to be achieved by renewable energy sources, respectively, by energy efficiency and other ways of carbon emission reduction. If consumption of total uranium resources, as estimated in 2009, was required to achieve a serious contribution of nuclear energy to carbon emission control by 2065, should one then conclude that nuclear energy cannot continue in production of carbon free energy with the same share in total energy production? This is question certainly very relevant for judgment on sufficiency of uranium resources and we try to answer it in Section 4. To obtain a quantitative base for this we continued our scenarios 2, 3, and 4 from the year 2065 up to 2100 on the power levels reached by the year 2065, i.e. with powers of 4878 GW, 3479 GW, and 2195 GW for strategies 2, 3, and 4, respectively. In view of the expected slow growth of total energy consumption in the last decades of the century the contributions of all three strategies to carbon emission reduction will remain substantial, not much below their values in 2065. For all three strategies we have calculated cumulative uranium requirements from 2010 through to 2100 without reprocessing and with reprocessing after 2065. Assumption of study was to postpone fuel reprocessing as late as 2065 in order to give sufficient time for development of all political, institutional and technical condition for safe use of plutonium. The required quantities of uranium without reprocessing are 49.8 Mt, 40.2 Mt, and 26.5 Mt for strategies 2, 3, and 4, respectively. The required quantities of uranium with reprocessing after 2065 are 44.1 Mt, 34.3 Mt, and 23.4 Mt for strategies 2, 3, and 4, respectively. The estimated uranium requirements until 2100 are upper limits as they are obtained by conservative assumption on the efficiency of uranium use, i.e. by assuming operation of present technology conventional reactors.

Even for the highest nuclear capacity growth of scenario 2 the uranium requirements are less than 50 Mt, the uranium resources estimated by simple crustal model.

For scenario 3, assuming plutonium recycle after 2065, the conservative estimate, based on the use of conventional reactors and ignoring reductions by the more efficient Generation 4 reactors, ends with uranium requirements on the level of 35.6 million tonnes up to the year 2100. In other words, keeping the present proven reactor technology, with plutonium recycle postponed to 2065, one could go on with a nuclear share of about one third in the total energy production until 2100 with approximately double uranium resources as estimated in 2009. Our figure without reprocessing until 2100 is about 13% higher and it amounts 40.2 Mt. While we can expect the conditions for reprocessing to exist by 2065, we can say that even the postponement of reprocessing until 2100 for strategy 3 with a very large contribution of carbon free energy results in still acceptable requirements. This is certainly so for the intermediate Strategy 4, which still contributes with about one quarter to required emission reduction, while the uranium requirements are lower.

Whether the introduction of reprocessing after 2065 will be necessary will depend on many future developments, such as the improvement of conventional nuclear technology, progress in fusion and CCS technology, rate of deployment of renewable resources, and of course, on the rate of increase of uranium resources. About this we cannot speculate. Also, we do not want to discuss in this place the wisdom or the feasibility of giving up nuclear energy in view of the enormous tasks world is facing to control the climate changes by GHG emissions. What we do want to show is that until the end of century uranium resources are not a limiting factor for a large nuclear contribution on the level of 3479 GW approximately, i.e. on the level of one third of total energy production, without introduction of such technologies as fast breeder reactors. That should be sufficient for a reasonable assurance that a strategy such as WEO 450 could be achieved, provided, of course, that renewable source and other ways of GHG emission control contribute their large shares. After 2065 there could be a welcome contribution from CCS installation, and, less likely, from fusion. If these developments fail, our estimates show that continued share of nuclear energy could be supported by conventional reactor technologies up to the end of century. Large scale introduction of fast breeders after 2100 would make the issue of uranium or thorium resources irrelevant for future energy production. Needless to say, in that case the uranium from the seawater would open as economically acceptable and for all practical purposes inexhaustible uranium source.

However, we do not want to overplay these future possibilities. It is not enough to show that nuclear energy is sustainable. This is easily done by assuming an early introduction of breeder reactors. However, in democratic societies nuclear energy must also be acceptable to most citizens. Nuclear energy must prove itself to be evidently safe, technically and politically. That is why it would be preferable to continue with proven technology till about the end of century. We show that possible from the point of resources. Many safety improvements were applied on the light water reactors after the Three Mile Island accident in 1979. There will be some lessons after Fukushima 2011 accidents. Applied, they will contribute further to the safety of present reactor line. Rather than changing basic technology too soon, it may be wiser to demonstrate several decades of safe and reliable operation of present one. That would be a good preparation for later introduction of new technologies, such as breeders. This is not a long delay, considering that with new technologies to use U238 and Th232 nuclear energy can serve humanity for thousands of years.

## 6. Conclusions

Under the long term sustainability of nuclear resources we understand the capability to support long term large share of nuclear energy (of about one third) in total energy production and in reduction of carbon emission. We determined the uranium requirement for corresponding nuclear strategies to 2065 and to the end of century. In view of our survey of non-conventional uranium resources with potential to substantially expand conventional uranium resources, as well as expected increase of conventional resources estimates relative to their 2009 values, and looking at the results of above presented nuclear strategies 2,3 and 4, we feel justified to conclude that, after nuclear build-up in the period 2025-2065, nuclear energy share on the achieved level of about 3479 GW, respectively about one third in the total energy production, can be sustained until the end of century using only proven conventional reactor technology or with introduction of plutonium recycle after 2065. Our conservative estimate indicate, that in later case about 35.6 million tonnes of uranium would be required by 2100 in that case. Postponing the spent fuel reprocessing until the end of century would increase uranium requirement to about 40.2 million tonnes.

Technologies and methods for improvement of nuclear fuel utilization have been considered. Even though some of these technologies are developed and proven technically viable, substantial implementations of these technologies are not expected in this century. While some effects on reduction of uranium requirements before the end of century may be possible, our aim for conservative estimates does not take them into account.

Looking to the end of century we note that based on a geochemistry model the total amount of uranium recoverable at price of 180 USD/kg U is estimated to 50 million tonnes.

On the technology side, large scale introduction of fast breeders after 2100 would make the issue of uranium or thorium resources irrelevant for future energy production.

Shorter and long term sustainability potential of nuclear fuel resources is enhanced by expected extraction of uranium from phosphates and seawater.

Finally, it may be concluded that nuclear fuel resources will not be a constraint for long term nuclear power development, even if the use of nuclear power is aggressively expanded.

## 7. References

Barthel, F.H. (2007). Thorium and Unconventional Uranium Resources, *Proceedings of a Technical Meeting „Fissile Materials Management Strategies for Sustainable Nuclear Energy"*, ISBN 92–0–115506–9, Vienna, Austria, September 2005.

Bodansky, D. (2004). *Nuclear Energy Principles, Practices and Prospects*, (Second Edition), Springer, ISBN 978-0387-20778-0, New York, USA

Bowman, C. D., Arthur, E. D., Lisowski, P. W., Lawrence, G. P., Jensen, R. J., Anderson, J. L., Blind, B., Cappiello, M., Davidson, J. W., England, T. R., Engel, L. N., Haight, R. C., Hughes, H. G., Ireland, J. R., Krakowski, R. A., Labauve, R. J., Letellier, B. C., Perry, R. T., Russell, G. J., Staudhammer, K. P., Versamis, G. & Wilson, W. B. (1992) Nuclear energy generation and waste transmutation using an accelerator-driven intense thermal neutron source, *Nuclear Instruments and Methods in Physics Research Section A*, Vol. 320, No 1-2, pp. 336-367, ISSN 0168-9002

Briggs, R.B. (1967) Summary of the Objectives, the Design, and a Program of Development of Molten-Salt Breeder Reactors, ORNL-TM-1851, Oak Ridge National Laboratory, USA

Cochran, T.B., Feiveson, H.A., Patterson, W., Pshakin, G., Ramana, M.V., Schneider, M., Suzuki, T. & von Hippel, F. (2010). *Fast Breeder Reactor Programs: History and Status*, International Panel on Fissile Materials, ISBN 978-0-9819275-6-5, Princeton, USA

Deffeyes, K.S. & MacGregor, I.D. (1980). World Uranium Resources, Scientific American, Vol. 242, No. 1, pp. 66-76, ISSN 0036-8733

Electric Power Research Institute [EPRI] (2009). *Nuclear Fuel Cycle Cost Comparison Between Once-Through and Plutonium Single-Recycling in Pressurized Water Reactors*, EPRI Report 1018575, Palo alto, California, USA

Ellis, T., Petroski, R., Hejzlar, P., Zimmerman, G., McAlees, D., Whitmer, C., Touran, N., Hejzlar, J., Weaver, K., Walter, J., McWhirter, J., Alhfeld, C., Burke, T., Odedra, A., Hyde, R., Gilleland, J., Ishikawa, Y., Wood, L., Myrvold, N., Gates III, W. (2010), Traveling-Wave Reactors: A Truly Sustainable and Full-Scale Resource for Global Energy Needs, 24.08.2011, Available from: http://icapp.ans.org/icapp10/program/abstracts/10189.pdf

Feinberg, S.M. (1958). Discussion Comment, *Rec. of Proc. Session B-10, ICPUAE*, United Nations, Geneva, Switzerland

Fetter, S. (2009), How long will the world's uranium supplies last?, 24.08.2011, Available from: http://www.scientificamerican.com/article.cfm?id=how-long-will-global-uranium-deposits-last

Fraser, J S., Hoffmann, C R., Schriber, S. O., Garvey, P M. & Townes, B M. (1981). A review of prospects for an accelerator breeder, Report AECL-7260, Atomic Energy of Canada Limited, Chalk River, Ontario, Canada

Forsberg, C.W., Renault, C., Le Brun, C., Merle-Lucotte, E. & Ignatiev, V. (2007), Liquid Salt Applications and Molten Salt Reactors, *Proceedings of ICAPP'07*, ISBN 9781604238716, Nice, France, May 13-18.

Garwin, R.L. (1998). The Nuclear Fuel Cycle: Does Reprocessing Make Sense?, *Proceedings of the Peer Review Workshop of the Pugwash Conferences on Science and World Affairs on the Prospects of Nuclear Energy: Nuclear Energy - Promise or Peril?*, ISBN 9789810240110, Paris, France, Dec. 1998

Gen. IV International Forum (2011a), Lead-Cooled Fast Reactor, 24.08.2011, Available from: http://www.gen-4.org/Technology/systems/lfr.htm

Gen. IV International Forum (2011b), Molten Salt Reactor, 24.08.2011, Available from: http://www.gen-4.org/Technology/systems/msr.htm

Heckrotte, W. (1977). Nuclear Processes Involved in Electronuclear Breeding, ERDA Information Meeting on Accelerator Breeding, BNL.

International Energy Agency [IEA] (2009). *World Energy Outlook 2009*, Organization for Economic (OECD), ISBN 978-9264061309, Paris, France

Intergovernmental Panel on Climate Change [IPCC] (2007). *Climate Change 2007 Synthesis Report*, IPCC, ISBN 92-9169-122-4, Geneva, Switzerland

Knapp, V., Pevec, D. & Matijević, M. (2010). The potential of fission nuclear power in resolving global climate change under the constraints of nuclear fuel resources and once-through fuel cycles, *Energy Policy*, Vol. 38, pp. 6793-6803, ISSN 0301-4215

Kouts, H.J.C. & Steinberg, M. (1977). Proceedings of an Information Meeting on Accelerator Breeding, *Conf-770107*, Brookhaven National Laboratory, Upton, New York 11973, January 18-19 1977.

Lewis, W.B. (1969). The Intense Neutron Generator and Future Factory Type Ion Accelerators, *IEEE Transactions on Nuclear Science*, Vol. 16, No 1, pp. 28-35, ISSN 0018-9499

Maniscalco, J.A., Berwald, D.H., Campbell, R.B., Moir, R.W. & Lee, J.D. (1981). Recent Progress in Fusion-Fission Reactor Design Studies, Nuclear Technology/Fusion, Vol. 1, No 4, pp. 419-478, ISSN 0272-3921

Massachusetts Institute of Technology [MIT] (2009). Update of the MIT 2003 Future of Nuclear Power, An interdisciplinary MIT Study

Nifenecker, H., Heuer, D., Loiseaux, J.M., Meplan, O., Nuttin, A., David, S. & Martin, J.M. (2003). Scenarios with an intensive contribution of nuclear energy to the world energy supply. *International Journal of Global Energy* Issues, Vol. 19, No 1, pp. 63-77, ISSN 0954-7118

Organization for Economic Co-operation and Development Nuclear Energy Agency [OECD/NEA] (2006). *Forty Years of Uranium Resources, Production and Demand in Perspective – The Red Book Retrospective*, OECD Publications, Paris, France

Organization for Economic Co-operation and Development Nuclear Energy Agency [OECD/NEA] & International Atomic Energy Agency [IAEA] (2010). *Uranium 2009: Resources, Production and Demand*, OECD Publications, ISBN 978-92-64-04789-1, Paris, France

Pevec, D., Knapp, V. & Matijevic, M. (2008). Sufficiency of the nuclear fuel, *Proceedings of the 7th International Conference on "Nuclear Option in Countries with Small and Medium Electricity Grids"*, ISBN 978-953-55224-0-9,Dubrovnik, Croatia, May 2008.

Rosenthal, M.W., Kasten, P.R. & Briggs, R.B. (1970). Molten-Salt Reactors—History, Status, And Potential, Nuclear Applications and Technology, Vol. 8, pp. 107-117, ISSN 0550-3043

Schneider, E.A. & Sailor, W.C. (2008). Long-Term Uranium Supply Estimates, Nuclear Technology, Vol. 162, pp. 379-387, ISSN 0029-5450

Sekimoto, H., Nagata, A. & Mingyu, Y. (2008). Innovative Energy Planning and Nuclear Option Using CANDLE Reactors, *Proceedings of the 7th International Conference on "Nuclear Option in Countries with Small and Medium Electricity Grids"*, ISBN 978-953-55224-0-9,Dubrovnik, Croatia, May 2008.

Steinberg, M., Grand, P., Takahashi, H., Powell, J.R. & Kouts, H.J. (1983). *The spallator - A new option for nuclear power*. Brookhaven National Laboratory report BNL 33020.

Tamada, M., Seko, N., Kasai, N. & Shimizu, T. (2006). Cost Estimation of Uranium Recovery from Seawater with System of Braid Type Adsorbent, Transactions of the Atomic Energy Society of Japan, Vol. 5, No 4, pp. 358-363, ISSN 1347-2879

United Nations Sigma XI Scientific Expert Group on Climate Change (2007). *Confronting Climate Change, 2007: Avoiding the Unmanageable and Managing Avoidable*, 27.07.2011., Available from: http://www.sigmaxi.org/programs/unseg/Full_Report.pdf

United States Department of Energy [USDOE] (2002). Generation-IV Roadmap: Report from Fuel Cycle Crosscut Group, 24.08.2011, Available from: http://www.ne.doe.gov/neac/neacPDFs/GenIVRoadmapFCCG.pdf

United States Nuclear Regulatory Commission (31.03.2011), History, 24.08.2011, Available from: http://www.nrc.gov/about-nrc/emerg-preparedness/history.html

World Information Service of Energy [WISE], (21.10.2010), Uranium Recovery from Phosphates, 24.08.2011, Available from: http://www.wise-uranium.org/purec.html

World Nuclear Association [WNA], (June 2011a), Advanced Nuclear Power Reactors, 24.08.2011, Available from: http://www.world-nuclear.org/info/inf08.html

World Nuclear Association [WNA], (June 2011b), Uranium from Phosphates, 24.08.2011, Available from: http://www.world-nuclear.org/info/phosphates_inf124.html

World Nuclear Association [WNA], (August 2011), Fast Neutron Reactors, 24.08.2011, Available from: http://www.world-nuclear.org/info/inf98.html

World Nuclear Association [WNA], (September 2011), Supply of Uranium, 24.08.2011, Available from: http://www.world-nuclear.org/info/inf75.html

# 2

# Radiation Induced Corrosion of Nuclear Fuel and Materials

Václav Čuba, Viliam Múčka and Milan Pospíšil
*Czech Technical University in Prague,*
*Faculty of Nuclear Sciences and Physical Engineering, Prague*
*Czech Republic*

## 1. Introduction

In some cases, a corrosion system may be exposed to radiation field. Most usually, it occurs as a part of atmospheric corrosion, when materials are, besides other factors, exposed also to visible and ultraviolet radiation. Technical significance of corrosion caused by such types of radiation is rather small. On the other hand, metallic materials may be exposed to high energy ionizing radiation (IR), emitted e.g. during operation of nuclear reactors, or during storage/processing of irradiated nuclear fuel. Contrary to non-ionizing radiation, IR significantly affects properties of corrosion system and reactions which proceed inside the system. Therefore, three main areas of research of radiation induced corrosion exist:

- corrosion of both fresh and irradiated nuclear fuel;
- corrosion of barrier materials (cladding of fuel elements, containers for irradiated fuel);
- corrosion of metal parts of nuclear and irradiation devices.

In the case of corrosion processes induced by IR (radiation oxidation), radiation-induced redox reactions are essential. Water plays very important role in all parts of nuclear fuel cycle - it is used as coolant and/or moderator in nuclear reactors; it is also used as coolant and/or shielding material in various radiation sources; ingress of deoxygenated granitic water into deep repository with spent nuclear fuel is an expected event. The direct effects of IR on corrosion of solid materials are usually negligible; in anoxic/deoxygenated aqueous environment, the majority of radiation damage and corrosion processes occur due to reactions of materials with products of water radiolysis. Therefore, strong influence of IR on corrosion processes may be expected in aqueous environment especially at higher temperatures. Radiation may change physico-chemical condition in corrosion systems, as well as its composition. Actions of primary intermediates of water radiolysis may complete change reactions proceeding in corrosion system (e.g. inside the container with irradiated nuclear fuel) - especially, they may increase electrochemical potential (Eh) and decrease pH of environment. Such changes may result in change of corrosion rate when compared to non-irradiated system. Also, due to irradiation, solid corrosion products of irradiated material may be formed.

This chapter is focused on review of radiation corrosion of uranium oxides and various materials used for nuclear waste packaging, shielding, or engineered barriers against migration of radionuclides from irradiated nuclear fuel matrix.

## 2. Radiolysis of water

As was mentioned above, radiolysis of water plays crucial role in corrosive processes occurring at the interface of liquid/solid environment in the presence of IR. Because of its optimum properties and availability, water is the most wide spread coolant used in nuclear reactors (Pikaev et al., 1988; Woods & Pikaev, 1994). It acts as a corrosive agent in the field of IR (namely at the presence of oxygen). Therefore, the development of nuclear powers depends also on the knowledge of the radiation chemistry of water. Also owing to this reason, water was among the first systems whose radiation chemistry was studied (Wishart & Nocera, 1998). But although our knowledge of the radiation chemistry of water at the steady state (when the concentrations of radiolytic products seems to be time-independent in the course of irradiation) at ambient temperature is rather complete, there are still not so many papers (Sehested & Christensen, 1987) on water radiolysis at the initial state of the process or at higher - up to supercritical - temperatures and pressures. Moreover, except for nuclear technology, the radiation-chemical transformations in liquid water are important also for radiobiological and medicine research and for "pure" radiation chemistry (Stepanov & Byakov, 2005). It is so because the water is, *inter alia*, a convenient object for investigating the key regularities of the interaction of IR with condensed matter (Erskov & Gordeev, 2008). Therefore, the basic features and mechanisms of the water radiolysis are contained in every monograph or general article devoted to radiation chemistry (Allen, 1961; Draganic & Draganic, 1971; Pikaev, 1986; Buxton, 1987; Spinks & Woods, 1990; Woods & Pikaev, 1994; Wishart & Rao, 2010).

### 2.1 Radiolysis of pure water

A general concept of radiolysis is conventionally divided into three main steps: the physical stage, physico-chemical stage and the chemical stage.

In the physical stage, the energy of ionizing particles, e.g. gamma photon or a charged particles, such as an electron, proton or an α-particle is transferred to water molecule which is ionized or excited to upper electronic states (usually denoted by a star). The products of these fast processes (less than $10^{-14}$ s) in the water are: $H_2O^+$, $e^-$, $H_2O^*$, $H_2O^{**}$ and $(H_2O^-)^*$. In the next physico-chemical stage ($10^{-14}$ - $10^{-12}$ s) the products undergo intra track ion-molecule reactions followed by further reactions such as fast dissociations, ionizations and recombination. Following products may be detected at this stage: H (0.62), OH (5.6), O, $H_2$ (0.15), $e^-_{aq}$ (4.78), $H^+_{aq}$ (4.78), $H_3O^+$, $H_2O^-$, $H_2O^*$. The yields at this stage are defined as the initial radiation chemical yields ($G_i^0$) and they are shown here, in some cases, in the brackets according to literature (Buxton, 1987) in the $10^{-2}\,eV^{-1}$ units (the radiation chemical yields - G-values - may be expressed as the number of species created/destroyed per 100 eV of absorbed energy or less often in SI units mol $J^{-1}$ which is equivalent to $9.65x10^6$ molecules x $10^{-2}eV^{-1}$). The initial yields were found (Watanabe & Saito, 2001) to be decreasing as the initial energy of monoenergetic electrons becomes low (in the energy region lower than 1 keV). For the third stage (chemical stage), it is typical that the resulting above mentioned transient species as well as new products of further ionizations are distributed along the track. The length of the track determines the number of primary events; e.g., for 100 eV electron, only about 10 primary events occur in water (Watanabe & Saito, 2001) along its 4 nm long track and for the 1 keV electron the highest density of chemical species is given for a radius of 20 nm. This leads to the formation of clusters of ions and radicals which are called spurs and blobs. The reactions proceeding within the spur structures in the time range from $10^{-12}$ to $10^{-7}$ s including their rate constants are given (Buxton, 1987; Motl, 2004) in Table 1. The two last reactions from the Table 1 take place just only in the case of high LET (linear energy transfer) radiation, e.g. α-particles or heavy accelerated ions. It is evident from the Table 1 that at the end of chemical stage there are the $OH^-$, $H_2O_2$ and eventually $HO_2$ species present in the spur (in addition to the above mentioned species). Overall, following are the species remaining when all the spur reactions are complete (Buxton, 1987): $e^-_{aq}$, H, OH, $HO_2$, $H_2O_2$ and $H_3O^+$. The radiation chemical yield of the product created in chemical stage is so-called primary yield usually denoted as $G_i$ or g(i). The primary yields depend, besides the presence of oxygen in water (Watanabe & Saito, 2001), strongly on the LET of applied radiation as is shown in Table 2 for the most important species from the point of view of material corrosion.

| **Reactions** | **kx10$^{-10}$** [L mol$^{-1}$ s$^{-1}$] |
|---|---|
| $e^-_{aq} + e^-_{aq} + 2H_2O \rightarrow H_2 + 2OH^-$ | 0.54 |
| $e^-_{aq} + OH \rightarrow OH^-$ | 3.0 |
| $e^-_{aq} + H_3O^+ \rightarrow H + H_2O$ | 2.3 |
| $H + H \rightarrow H_2$ | 1.3 |
| $OH + OH \rightarrow H_2O_2$ | 0.53 |
| $OH + H \rightarrow H_2O$ | 3.2 |
| $H_3O^+ + OH^- \rightarrow 2H_2O$ | 14.3 |
| $OH + H_2O_2 \rightarrow HO_2 + H_2O$ | 0.0045 |
| $O + OH \rightarrow HO_2$ | - |

Table 1. Spur reactions in water

| **LET** [eV nm$^{-1}$] | **G** [$10^{-2}$eV$^{-1}$] | | | | | |
|---|---|---|---|---|---|---|
| | **$e^-_{aq}$** | **OH** | **H** | **$HO_2$** | **$H_2$** | **$H_2O_2$** |
| 0.23 | 2.63 | 2.72 | 0.55 | 0.008 | 0.45 | 0.68 |
| 12.3 | 1.48 | 1.78 | 0.62 | - | 0.68 | 0.84 |
| 61 | 0.72 | 0.91 | 0.42 | 0.05 | 0.96 | 1.00 |
| 108 | 0.42 | 0.54 | 0.27 | 0.07 | 1.11 | 1.08 |

Table 2. Dependence of the primary yields of some radiolysis products on LET (Buxton, 1987)

The comparison of the values from the Table 2 with those presented earlier for the initial yields gives evidence that the $G_i^0$ values are quite different from the $G_i$ values. It is worth mentioning that the experimentally measured yields at a stationary state (G(i)) are also different both from the $G_i^0$ and $G_i$ values. Besides both the mechanism of reactions and radiation chemical yields, water radiolysis is characterized also by material balance equations. For the primary products, following equation applies (Buxton, 1987):

$$G_{-H2O} = 2G_{H2} + G_H + G_{e-(aq)} - G_{HO2} = 2G_{H2O2} + G_{OH} + 2G_{HO2} \quad (1)$$

(for a low LET radiation without the $G_{HO2}$ or $2G_{HO2}$ terms). Next, at the time greater than $10^{-7}$ s the radiolysis products begin to diffuse randomly and either react together or escape into the bulk solution. The reactions at this time period proceed outside the spurs. They lead, *inter alia,* to the creation of oxygen *via* decomposition of perhydroxyl (hydroperoxide) radicals $HO_2$ as well as to the reverse conversion both of hydrogen and hydrogen peroxide back to the molecules of water and H or OH radicals or various secondary radicals according to the oxidizing or reducing conditions in water. In such a way, the final stable products of water radiolysis are hydrogen, hydrogen peroxide and eventually oxygen. However, the decomposition of pure water is minimal, including in pressurized-water reactors (PWRs) (Wishart & Rao, 2010), owing to the back reactions which re-form water from the above mentioned stable products (Spinks & Woods, 1990). The charged particles created in radiolytic process are hydrated in the water or solvated in other solvents.

It was found (Wishart & Rao, 2010) that the H bond network of water needs approximately 1.6 ps to accept a new created "dry" electrons $e^-$ and transform them to the "wet" $e^-_{aq}$ electrons. Water molecules create a hydration shell around the electron forming, in such a way, cavities surviving some tens of fs ($10^{-15}$ s). The standard reduction potentials of hydrated electron range from -2.77 V (Wishart & Nocera, 1998) to -2.9 V (Buxton, 1987). Its half-life reaches about of $2.1 \times 10^{-4}$ s in neutral water. Hence, it is powerful reducing agent acting in one-electron transfer processes. The electrons are assumed to be transported in water medium until their energies fall below 7.4 eV (the threshold for electronic excitation in liquid water (Watanabe & Saito, 2001)). Using the ultrafast liquid-jet photoelectron spectroscopy measurements, it was possible to postulate (Siefermann et al., 2010) the existence of hydrated electron at the water surface with the lifetimes longer than 100 ps. Its characteristic may be different from the electron hydrated in the bulk solution.

The OH radical, besides the molecular hydrogen peroxide, as the most important oxidizing species (the standard redox potential $E^0(OH/OH^-)$ is 2.32 V at pH = 7), is assumed to be closely related to the corrosion of the materials in nuclear reactors as well as the containers containing the spent nuclear fuel. Other oxidants such as hydroperoxide $HO_2$ and molecular oxygen are produced in very low quantities with low LET radiations. The OH radical often reacts *via* a simple electron transfer ($R^- + OH \rightarrow R^{\cdot} + OH^-$). In a strongly alkaline solution, OH is rapidly deprotonized with $OH^-$ ions producing a new $O^-$ radical with the half-life of 5 ns (Wishart & Nocera, 1998):

$$OH + OH^- \rightarrow O^- + H_2O. \quad (2)$$

The molecular hydrogen can transform the oxidizing OH radical to the reducing H radical which can be transformed up to $e^-_{aq}$ under severe (high pH and pressure of 10 MPa) conditions:

$$OH + H_2 \rightarrow H + H_2O; \quad (3)$$

$$H + OH^- \rightarrow H_2O + e^-_{aq}. \quad (4)$$

As the Monte Carlo simulations showed (Watanabe & Saito, 2001), more reactions of OH radicals occur for 1 and 10 keV electrons than for 1 MeV electrons. The recombination of two OH radicals leads fast to creation of hydrogen peroxide (Table 1). That is, why the decay of OH radicals occurs on the same time scale as formation of $H_2O_2$ molecules. But as it was shown by La Verne (La Verne, 2000), not all of the OH radicals lead to production of hydrogen peroxide.

The H atoms are connected with the hydrated electrons which may be assumed to be conjugate base of H radicals. Hence, the H atoms create very fast ($k=2.3x10^{10}$ L $mol^{-1}$ $s^{-1}$) (Wishart & Nocera, 1998) from $e^-_{aq}$ in acidic solutions: $e^-_{aq}+H^+_{aq}\rightarrow H$. The reverse conversion of H radicals back to the $e^-_{aq}$ is possible in highly alkaline solution but this reaction is about three orders slower than the mentioned last one. The standard reduction potential of H atoms equals -2.1 V (that is a similar value as for the $e^-_{aq}$ – see above).

The perhydroxyl radical (or hydroperoxide) $HO_2$ is a secondary radical formed in oxygenated solutions from primary radicals such as:

$$H + O_2 \rightarrow HO_2; \quad (5)$$

$$O_2 + e^-_{aq} \rightarrow O_2^-; \quad (6)$$

$$O_2^- + H^+ \leftrightarrow HO_2. \quad (7)$$

As it can be seen from the last reaction, the $HO_2$ forms in acidic solutions from the deprotonized $O_2^-$ with pK of $HO_2$ equal to 4.7 (Buxton, 1987). The standard reduction potentials are -0.05 V for $HO_2$ (acidic solutions) and -0.33 V for a stronger reducing agent $O_2^-$. On the other hand, the $HO_2$ radical is a stronger oxidant than is the $O_2^-$.

From the text above it is clear that the features of water radiolysis depend on conditions of the radiolysis. First of all, the primary yields of $e^-_{aq}$, H, OH, $H_2O_2$ (not $H_2$) decrease (Buxton, 1987) with increasing pH but only up to pH = 3. In the range from pH = 3 up to pH = 13, the $G_i$ values seem to be unchanged. But, especially in the alkali region "the effect of pH on the primary yields is still an open question" (Ferradini & Jay-Gerin, 2000). The increasing pH value leads (Pikaev, 1986) also to the decreasing in stationary concentrations of molecular hydrogen and hydrogen peroxide. Another important factor affecting both stationary concentration of molecular products and the $G_i$ values is the LET. It is well known (Ferradini & Jay-Gerin, 2000) that the first mentioned quantity increases with increasing LET. In the same manner change the primary yields of molecular products and $HO_2$ radical but, on the other hand, the G-values of primary radicals decline in this direction. The effect of pH and LET on the $G_i^0$ quantity was found to be similar as in the case of the primary $G_i$ yield. The dose rate may affect the G values or the stationary concentrations in the same manner as the LET does.

Although the main characteristic data on water radiolysis are independent on the temperature up to about 100°C, their changes at higher temperatures are conspicuous and important e.g. for the gas evolution from water heat-transfer agent. Unfortunately, there is still only few and often conflicting data on the water radiolysis at the higher temperatures and pressures. Many new reactions with high activation energies are suggested (Sehested & Christensen, 1987) at the temperatures 200-300°C (e.g. $O_2^- + O_2^- \leftrightarrow O_4^{2-}$; $O_4^{2-} + H^+ \rightarrow HO_2^- + O_2$; $O_3^- + O^- \rightarrow O_4^{2-} \rightarrow 2O_2^-$). As a rule, the yield of water radiolysis ($G(-H_2O)$) and the yields of primary radical products ($G_i$) increase and the yields of molecular products decrease with increasing temperatures, though some experimental data (e.g. the yield of hydrated electrons (Sehested & Christensen, 1987)) may be contradictory. These all changes lead to an increase in the stationary concentration of hydrogen in an open system (the hydrogen gas can escape from the system) (Pikaev, 1986). They are connected (Pikaev et al., 1988) with a broadening of spurs and creation of more diffusive distribution of primary products at higher temperatures. At higher temperatures, the probability of recombination processes of primary products in spurs is lower than at ambient temperatures. The radiolysis of water is extremely affected by the temperature especially in the supercritical state ($T > 374$°C; $P > 22.1$ MPa). The study of such systems is motivated by the development of the 4$^{th}$ generation of nuclear reactors using supercritical water as the coolant. The density of supercritical water is much lower than that under the ambient conditions and, therefore, the hydration shell is composed from only a small count of molecules of water. That is, besides other things, the reason why different radiation chemical yields (and their ratios), the reaction rate constants and many spectral properties of transient species were observed (Wishart & Rao, 2010) under supercritical conditions in comparison with the normal conditions. Besides, the Arrhenius law does not apply for many radical reactions in the supercritical state. It was experimentally proved that, under these conditions, the radiation yields $G(e^-_{aq})$, $G(OH)$ and the sum ($G(e^-_{aq}) + G(OH) + G(H)$) dramatically increase with increasing temperature in the supercritical range. Therefore, all simulations of a radiolytic processes should consider this anomaly.

A deeper insight into the radiolytic processes in liquid water provides pulse radiolysis, leading to the detailed knowledge of the ultrafast primary events within the spurs. After the nanosecond pulse radiolysis, the picosecond pulse accelerated electron beams were used (Belloni et al., 2005) from the year 1968. Nowadays, the time resolved ultra-short pulses in the femtosecond (1 fs=$10^{-15}$ s) and, in near future, attosecond (1 as=$10^{-18}$ s) regime may be used (Haarlamert & Zacharias, 2009). Using the excellent equipments working in the above mentioned regime it was for example shown (Crowell et al., 2005) that in the time interval below 5 ps after ultrafast laser photo-ionization there are many complicated events between electronic relaxations and vibrational energy redistribution. But it is worth mentioning that the IR can not be adequately replaced by laser irradiation owing to different mechanisms of energy deposition in the condensed phase. It was also found (Buxton, 1972) that the spur structures in irradiated water are practically complete at the time of 120 ns. By the same time, the primary radiation yield of hydrated electrons $G_{e\text{-}aq}$ was found to be 2.8 x $10^{-2}$ $eV^{-1}$ whereas at the time of 7.5 ns it reaches the value of 3.6 and the initial radiation yields $G_{e\text{-}aq}{}^0$ measured by the femtosecond technique was found to be of 4.6 ± 0.3 x $10^{-2}$ $eV^{-1}$ (Yang et al., 2011). Using the same fast pulse water radiolysis, it was discovered that about 10% of hydrated electrons react in spurs with the $H_3O^+$ and OH species within 10 ps. Similarly, the lifetime of the basic initial particle $H_2O^+$ was assumed to be less than 100 fs and the hydration time of the "dry" electron may be much lower than it was determined earlier (see above), i.e. of 250-500fs (Domae et al., 1996). The results obtained in the pulse radiolysis experiments including application of the concept of equivalent velocity spectroscopy (Yang et al., 2009), are often used for the simulation of the radiolytic reactions. Up to now, there were developed several kinds of various models such as widely known diffusion model, Freeman´s model or Hamill´s model. None of them describes the radiolytic process exactly. The quality of every one can be considered according to the agreement of the $G_i^0$ values calculated from the model with them following from the measurements or according the reaction of model on the changes in the LET values or on the presence of various solutes. From this point of view, the diffusion model seems to be most convenient. The model is assumed for all chemical species in the track to have the Gaussian distribution. For the better description, namely when the diffusion is slow or for the short times (sub-picosecond) or for a large number of particles, the improved extended spur diffusion models are formulated (Stepanov & Byakov, 2005; Domae et al., 1996). Mostly, they consider 32 or 34 basic elemental reactions (Stepanov & Byakov, 2005; Erskov & Gordeev, 2008; Watanabe & Saito, 2001; Domae et al., 1996) leading to the decomposition of water. The Laplace transform method is applicable in many cases. Using these methods, it was for example found that no evidence has been obtained to support the dry charge recombination (Buxton, 1972).

The pulse radiolysis of water and the simulation of radiation processes play an important role in the storage of nuclear waste or a spent nuclear fuel (Wishart & Rao, 2010). Generally, the radiolysis of water increases the rate of $UO_2$ dissolution (Christensen & Sunder, 2000) because the redox conditions may be changed due to radiolytic products and the stability of the waste form and container may be affected by them. The main role in these processes posses the oxidants (i.e. $H_2O_2$ and the alkaline form of $HO_2$ radical - $O_2^-$ radical). Their concentrations in oxygenated solutions increase with increasing total dose. The saturated concentration of $H_2O_2$ is assumed to be about of $10^{-3}$ mol dm$^{-3}$ (Sunder & Christensen, 1993). The suggested models of the corrosion processes are in agreement with measured rates in the case of gamma radiolysis and in unirradiated solutions containing oxygen or hydrogen peroxide. There are some difficulties with α-radiolysis or, generally, with a high-LET radiation. Therefore, further studies are required to fully understand the effects of high-LET radiation on corrosion of used nuclear fuel.

In connection with the storage of nuclear waste or spent fuel, a special attention is devoted to the radiolysis of water in micro- and nano-porous confining materials such as concrete and clays, zeolites, silica gels, etc. which trap important quantities of interstitial water. Many characteristics of water radiolysis may be changed in micro- or nano-confined water in comparison with the bulk process: real dose absorbed by water, kinetics, radiation chemical yields of various radiolytic products. These changes may be due to the high viscosity of water inside the pores, the formation of OH radicals *via* transfer of holes from the solid material to the water ($h^+ + H_2O \rightarrow H^+ + OH$) or creation of hydrated electron *via* transfer of electron from the ionized solid to the water ($e^- + H_2O \rightarrow e^-_{aq}$) and, last but not least, the energy transfer from a solid phase to the chemisorbed or physicosorbed molecules of water. As a consequence of these (and many others) processes, high radiation yields are found for one or two mater layers which are decreasing with increasing size of pores. Therefore, the decrease in reaction rates of $e^-_{aq}$ or OH with organics in the course of pulse radiolysis in zeolites was found (Liu et al., 1997).

Just as a matter of interest, it may be mentioned that according some authors (Sawasaki et al., 2003) the gamma radiolysis of water could serve in the future as a manner for the hydrogen gas production. They suggest to combine two methods: first-the conversion of gamma radiation emitted by radioactive waste to electrons and photons with energies from tens to thousands eV, using the high Z-materials (Ta, Pd and Pb) and secondly the removal of molecular hydrogen from the system, using hydrogen occluders (Pd, Ta). But, the efficiency of this method seems to be still very low. May be that a higher efficiency may be achieved when some solid promoters participating in the H-radical production will be used (Wishard & Rao, 2010).

In conclusion, it may be said that even though many aspects of the water radiolysis are very well known further systematic studies (both experimental and theoretic) are needed to fully understand both the early elemental processes of the radiolysis and the reactions under extreme conditions.

### 2.2 Radiolysis of aqueous solutions relevant to nuclear waste programs

The corrosion resistance of various parts of containers with spent nuclear fuel depends mainly on the geological subsoil and chemical composition of surrounding groundwater. (Nishimura, 2009). Both mentioned factors are different in various localities. Several concepts of repository for geological disposal such as rock salt, clay/argillaceous rock and crystalline (granitic) rock were compared according to different parameters (FMET, 2008) with result that the disposal site built in granitic structure appears to be most suitable for minimizing the entry of radionuclides into the biosphere. Therefore, nuclear fuel waste programmes by many countries, not only European, are planned for disposal vault in granitic rock (Sunder & Christensen, 1993). Because these sites are located some hundreds of meters underground, the concentration of oxygen dissolved in water is very low due to its consumption in various chemical and biochemical processes (Pitter, 1999).

The most likely pathway for releasing of corrosion products and radionuclides appears to be transport of groundwater namely its ingress into repository or to the inner container with latent material defect, although the dissolution rate of fuel, predominantly $UO_2$ in anoxic reduction environment is very low (Parks & Pohl, 1988). The redox conditions may be substantially changed due to radiolysis of water by IR associated with the fuel which produces not only reductive but also several oxidative species with high reactivity (Spinks & Woods, 1990). The last mentioned radiolytic products substantially increase the corrosion rate of all relevant materials and parts of container.

It is generally known that contrary to pure water the irradiation of diluted aqueous solutions of inorganic compounds present predominantly in dissociated forms leads to various chemical and physicochemical processes such as redox reaction, gas evolution, changes in pH value, formation of precipitate etc. All these processes appear to be a result of mutual interaction of dissolved compounds with products of water radiolysis, while the effect of direct interaction of radiation with dissolved compound is virtually negligible. On the contrary, with concentrated solutions (higher then 1 mol $dm^{-3}$) the direct deposition of radiation energy to the solute cannot be neglected. Energy deposition in this case occurs in proportional to the electron fraction of each components i. e. solvent ($H_2O$) and corresponding solute (Katsumura, 2001). In both cases the radiolysis and concentration of radiolytic products are affected by many another factors such as nature of radiation, dose rate, composition of solution undergoing radiolysis, pH value and presence of different gases especially oxygen dissolved in solution.

Various inorganic ions present in granitic groundwater (Ollila, 1992) such as $CO_3^{2-}$, $HCO_3^-$, $HSO_4^-$, $SO_4^{2-}$, $Cl^-$ and cations $Na^+$, $Mg^{2+}$, $Ca^{2+}$, $Fe^{2+}$, $Si^{4+}$ act as scavengers of intermediates of water radiolysis further affecting the properties and behavior of the whole system. With regards to corrosion resistivity of containers and its components, among aforementioned species are very important (hydro)carbonate ions $CO_3^{2-}$ and $HCO_3^-$ which are quite strong scavengers for OH radical (Buxton et al., 1988):

$$OH + CO_3^{2-} \rightarrow CO_3^- + OH^- \tag{8}$$

$$OH + HCO_3^- \rightarrow CO_3^- + H_2O. \tag{9}$$

Therefore, aqueous solutions of $Na_2CO_3$ saturated with inert gases such as Ar, He, $N_2$ are used for simulation of underground granitic water (Sunder and Christensen, 1993). After achieving the steady state during sufficient long period gamma radiolysis, the concentration of radiolytic products is relatively low. However created carbonate radicals appear to be strong oxidative agents, analogously to OH radicals. The mechanism of alkaline solution containing carbonate during irradiation with low LET includes further following elemental reactions:

$$O_2^- + CO_3^- \rightarrow CO_3^{2-} + O_2 \tag{10}$$

$$H_2O_2 + CO_3^- \rightarrow HCO_3^- + O_2^- + H^+ \tag{11}$$

$$HO_2^- + CO_3^- \rightarrow CO_3^{2-} + O_2^- + H^+ \tag{12}$$

$$H^+ + HCO_3^- \rightarrow CO_2 + H_2O \tag{13}$$

$$OH^- + HCO_3^- \rightarrow CO_3^{2-} + H_2O \tag{14}$$

$$H_2O + CO_2 \rightarrow HCO_3^- + H^+ \tag{15}$$

$$H_2O + CO_3^{2-} \rightarrow HCO_3^- + OH^- \tag{16}$$

$$H_2O + CO_4^{2-} \rightarrow HO_2^- + CO_2 + OH^- \tag{17}$$

$$CO_3^- + CO_3^- \rightarrow CO_4^{2-} + CO \tag{18}$$

Saturation with $N_2O$ leads to the increase of steady-state concentration of $H_2O_2$, and ionradical $O_2^-$ whereas concentration of oxidizing OH radical does not change:

$$e^-_{aq} + N_2O \rightarrow O^- + N_2 \tag{19}$$

$$O^- + H_2O \rightarrow OH + OH^- \tag{20}$$

$$OH + OH \rightarrow H_2O_2 \tag{21}$$

$$OH + H_2O_2 \rightarrow HO_2 + H_2O \tag{22}$$

$$HO_2 + HO_2 \rightarrow H_2O_2 + O_2^- \tag{23}$$

$$HO_2 \rightarrow H^+ + O_2^-. \tag{24}$$

Rather different conditions are in oxygenated solutions possessing a high concentration of $H_2O_2$ and oxidizing species $O_2^-$. Presence of oxygen promotes the formation of $O_2^-$ due to reaction:

$$e^-_{aq} + O_2 \rightarrow O_2^-. \tag{25}$$

On the contrary, the concentration of oxidizing OH radicals is relatively low due to their removal by reaction (22). Moreover the OH radicals may be intensively scavenged by addition of formate ions or t-butanol. The former case appears to be less suitable because the amount of $O_2^-$ radical formed by reaction (25). Its amount is increased (Sunder et al., 1992) due to reactions:

$$OH + HCOO^- \rightarrow CO_2^- + H_2O \tag{26}$$

$$CO_2^- + O_2 \rightarrow CO_2 + O_2^-. \tag{27}$$

As compared with carbonate or hydrocarbonate ions the reactivity of sulfate or hydrosulfate anions is rather different (Jiang et al., 1992). It is supposed that reaction of OH radicals with sulfate anion

$$OH + SO_4^{2-} \rightarrow SO_4^- + OH^- \tag{28}$$

leading to the sulfate radicals is very slow or does not occur at all. Therefore $HSO_4^-$ and undissociated $H_2SO_4$ react with OH radicals forming sulfate radical:

$$OH + HSO_4^- \rightarrow SO_4^- + H_2O \tag{29}$$

$$OH + H_2SO_4 \rightarrow HSO_4 + H_2O \rightarrow SO_4^- + H_3O^+. \tag{30}$$

Moreover, in concentrated sulfuric acid solution, the fast formation process of sulfate radicals takes place in the first step probably by the electron abstraction:

$$H_2SO_4 \rightarrow H_2SO_4^+ + e^-; \text{ or } HSO_4^- \rightarrow HSO_4 + e^-. \tag{31}$$

The decay processes include minimally five elemental reactions leading to molecular oxygen and peroxydisulfate $S_2O_8^{2-}$:

$$SO_4^- + SO_4^- \rightarrow S_2O_8^{2}. \tag{32}$$

It follows from the study (Anbar and Thomas, 1964) that OH radicals react with $Cl^-$ ions in the presence of $H^+$ to produce $Cl_2^-$:

$$OH + Cl^- + H_3O^+ \rightarrow Cl + 2H_2O \tag{33}$$

$$Cl + Cl^- \rightarrow Cl_2^-. \tag{34}$$

Therefore reaction (33) is strongly affected by pH and the formation of transient $Cl_2^-$ and its yield also depends on the pH and concentration of chloride ions. Hydroxyl ions may be formed e.g. by reaction

$$e^-_{aq} + H_2O_2 \rightarrow OH^- + OH. \quad (35)$$

Moreover it is supposed (Sunder and Christensen, 1993) that the formation of $HOCl^-$ takes place

$$OH + Cl^- \rightarrow HOCl^- \quad (36)$$

and that under suitable concentration of $Cl^-$ ions the significant percentage of the OH radicals would exist as $HOCl^-$.

The presence of $Fe^{2+}$ ions in groundwater may be caused mainly by its ingress into container or by leaching of packaging material, most commonly carbon steel. Due to anaerobic conditions the only oxidizing species are products of water radiolysis, predominantly OH and $O_2^-$ radicals which can oxidize ferrous ions e. g.

$$Fe^{2+} + OH \rightarrow Fe^{3+} + OH^- \quad (37)$$

Another species such as $H_2O_2$ and product of its radiation decomposition-$HO_2$ radicals are also strong oxidizing agents (Ferradini and Jay-Gerin, 2000; Daub et al., 2010). Moreover both $Fe^{2+}$ ions and formed ferric ions may be hydrolyzed

$$Fe^{2+} + H_2O \rightarrow Fe(OH)^+ + H^+ \quad (38)$$

$$Fe^{3+} + 3H_2O \rightarrow Fe(OH)_3 + 3H^+ \quad (39)$$

and deposited hydroxide may be converted to oxide:

$$2Fe(OH)_3 \rightarrow Fe_2O_3 + 3H_2O. \quad (40)$$

Sulfate radicals can also react with species present in granitic groundwater. In more detail e.g. the kinetics of their reactions with $Cl^-$ or $HCO_3^-$ ions was studied (Huie and Clifton, 1990) as well as with $Fe^{2+}$ ions (McElroy and Waygood, 1990), leading also to the oxidation $Fe^{2+} \rightarrow Fe^{3+}$ and to the decay of $SO_4^- \rightarrow SO_4^{2-}$.

Special group represent aqueous solutions of actinide ions and fission fragments during the reprocessing of spent nuclear fuels. They undergo different valence changes due to irradiation by numerous radionuclides. The understanding of elemental reactions taking place is very important especially for separation of individual elements. Therefore this problematic has been the subject of various studies (Pikaev et al., 1988; Vladimirova, 1990).

## 3. Radiolytically induced corrosion

Radiolytic corrosion of nuclear fuel and/or parts of nuclear devices presents significant and long term studied problem. To emulate corrosive processes in various cooling loops (used in nuclear reactors or irradiation devices), cooling pools (for storage of irradiated fuel) and deep repositories, the studies are performed in both pure water and aqueous solutions relevant to the anticipated conditions. Moreover, since significant amount of heat is often generated during interaction of IR with matter, the radiation-corrosion experiments are often performed at elevated temperatures; this makes such studies even more experimentally difficult. Rate of both non-radiolytic and radiolytic corrosion may be expressed as an increase of the corrosion products layer on corroded material per time unit.

Corrosion processes (and effects of radiation) are often evaluated via changes in corrosion potentials; hydrogen evolution due to corrosion may be also followed. Characterization of corrosion products may help in making quantitative assumptions regarding mechanisms of corrosion processes.

### 3.1 Nuclear fuel

Irradiated nuclear fuel is accumulated worldwide in many countries using nuclear power. Although the opinion that reprocessing of nuclear fuel is more convenient from both economical and environmental point of view dominates nowadays, concept of nuclear fuel disposal in corrosive resistant containers is being still carefully considered. Proper disposal method should ensure the release of the least possible amount of radionuclides back to geo- and biosphere. Currently, deep repositories are prepared for future use. Until that time, irradiated fuel is stored in transport/storage containers, placed in separate repositories, often directly in the areas of nuclear power plants.

Safety of storages and repositories for radioactive waste may be proven only by experimental studying and modelling of all processes that may lead to release of radionuclides to the environment. For more than 30 years, those processes are studied in many countries operating nuclear power plants. Thorough safety analyses, including studies of all anticipated processes and incidents that may occur during time of high radioactivity of irradiated fuel proved, that by disposal of irradiated nuclear fuel in dry corrosion resistant containers surrounded by compacted bentonite in the granite structure hundreds of meters below surface, it is possible to prevent releasing of radionuclides into geo- and biosphere above limits allowed by state authorities. It is presumed that the most probable carrier of corrosion products to biosphere would be deoxygenated groundwater, because its ingress into disposal site must be expected. Another factor, which cannot be totally neglected, is human random mistake, resulting in disposal of container with hidden defect. This may lead to damage of container in shorter time than is its projected lifetime (e.g. 1000 years in Czech reference project of deep repository). Consequently, groundwater may ingress to a container with irradiated nuclear fuel during still high activity of gamma and beta emitters. As was discussed in previous sections, the mechanism of corrosion of materials of containers and fuel elements cladding may change due to irradiation. Moreover, significant change of conditions inside the container may occur, as well as faster degradation of materials and faster oxidation of fuel matrix, presenting one of the most significant barriers against release of radionuclides. It was shown (Bruno et al., 1999), that radionuclides are firmly bonded in fuel matrix in the case, that oxidation state of $UO_2$ matrix does not exceed upper limit of stability, corresponding to stoichiometry $U_3O_7$.

Probability of such scenario is rather low and depends on level of quality control during radioactive waste disposal. According to data published on this topic (Andersson, 1999), we may expect that probability of defective container occurrence, which evades noticing of personnel is lower than $1 \times 10^{-3}$, i.e. 1 container of 1000. However, those containers may present the greatest risk for safety of whole disposal site. Study of influence of radiation on inner environment of container with spent nuclear fuel is relevant also for storage of spent nuclear fuel containing residual water after reloading of fuel from pools near reactor into transport/storage containers.

Direct research of environment inside storage container was performed by authors from USA (Domski, 2003) and Canada (Shoesmith et al., 2003). The first of both works is related to preparation of deep repository in Yucca Mountain, USA. This repository is placed in non-saturated zone, i.e. in oxidative environment. On the contrary, some other repositories (e.g. in Europe) are planned in granitic structures hundreds of meters below surface. Generally, corrosion may be described by following oxidative reaction

$$M \rightarrow M^{n+} + ne^- \quad (41)$$

where n is number of exchanged electrons. In closed environment of disposal container after water ingress may formed ions further hydrolyze for example by reactions (38) and (39), with simultaneous formation of hydrogen ions. Thus, hydrogen ions may accumulate inside the container, which leads to significant decrease of pH. Solubility of compounds forming protective films on surface of metals may increase due to low pH, which may lead to further accelerating of degradation of materials inside the storage container, including faster dissolution of spent nuclear fuel (Burns et al., 1983; Smart et al., 2008). In the work of (Pan et al., 2002) authors observed value of pH = 2 due to hydrolysis of inner metal parts of container.

Rate of processes occurring inside the container is also significantly affected by corrosion products precipitated on surface of metallic materials. Character of those products depends on material of container and also on composition of groundwater and on other components of disposal site (e.g. bentonite). Under anoxic reductive conditions, water accepts electrons released during anodic dissolution of metals and following reactions occur:

$$2H_2O + 2e^- \rightarrow H_2 + 2OH^- \text{ (alkaline environment)} \quad (42)$$

$$2H^+ + 2e^- \rightarrow H_2 \text{ (acidic environment)} \quad (43)$$

The rate of those reactions is very low and therefore, the rate of metal corrosion in anoxic environment is also very low. However, as was described in section 2.1, OH radicals are formed during water radiolysis. Such strong oxidants may significantly contribute to corrosion of metals (or metal ions) by their transfer to higher oxidation states:

$$M^{n+} + OH \rightarrow M^{n+1} + OH^- \quad (44)$$

Due to the fact, that groundwater will be presumably free of oxygen in the environment of geological disposal site, the only oxidizing species after groundwater ingress into geological disposal site will be reactive intermediates of water radiolysis. Reactive intermediates may increase rate of corrosion due to intensification of cathode depolarization process (Kaesche, 1985; Lillard et al., 2000). Concentration of compounds acting as strong cathodic reactants will increase due to radiation. However, hydrated electrons may reduce dissolved metal ions to lower valences or into metallic form (Buxton et al., 1988) according to equation

$$M^{n+} + e^-_{aq} \rightarrow M^{(n-1)+} \quad (45)$$

and thus decrease the corrosion rate.

Lots of works were published, dealing with experimental study, evaluation or modeling of radiolytic processes related to corrosion of uranium or nuclear fuel in water. A model, based on radiation energy deposition in the pore water and other constituents of natural uranium ore, was developed for radiation energy deposition and its consequences for water radiolysis (Liu & Neretnieks, 1996). Monte Carlo method was used for calculation of randomly generated radiation within grains of uranium minerals in the ore. The radiation energy was then allowed to deposit into the various constituents in the ore. The model was used for prediction of IR with different LET (alpha, beta, and gamma decay). The results showed that in the real system, due to dissipation of energy in solids, only a small fraction of total radiation energy may be available for radiolysis (and possible participating in corrosion processes). Generally, the efficiency of radiolysis (the actual oxidant production rate over the maximum possible oxidant production rate) is about 1%. (Shoesmith et al., 2003) used a mixed-potential model to predict fuel corrosion within a failed nuclear waste container under anticipated repository conditions. The model accounts for the effects of high LET radiation (alpha radiolysis), the formation of corrosion product deposits on both the fuel surface and the surface of the carbon steel container liner, and the homogeneous redox reactions between radiolysis products and soluble fuel and steel corrosion products. The model was used to predict the corrosion potential - usually defined as steady state corrosion potential in volts versus standard (e.g. saturated calomel) electrode - and corrosion rate of the fuel and the extent of fuel conversion over an exposure period of $10^6$ years.

Fuel matrix in contact with water constitutes a dynamic redox system due to the time dependent radiolytic generation of oxidants and reductants at the fuel interface. (Bruno et al., 1999) performed experimental determination and chemical modelling of radiolytic processes at the irradiated fuel/water interface to better understand the main processes and mechanisms that control the impact of radiolytically generated reactants on the stability of $UO_2$ matrix and release of radionuclides. Since the actual fuel rods fragments were used in the experiments, the corrosion system was exposed to a mixture of alpha, beta and gamma radiation. Mass balance calculations indicate that consumption of radiolytically produced oxidants by the fuel corresponds to the formation of and oxidized $UO_{2+x}$ surface layer in distilled water and the formation and release of soluble U(VI) carbonate complexes in bicarbonate media. It was concluded, that uranium release in early contact times is controlled by oxidative dissolution of the fuel matrix. This process also controls the release of Sr, Np and Pu. The measured concentrations of actinides appeared to be limited by the solubility of Ac(IV) hydroxide phases. The release of Tc and Mo appeared to be controlled by oxidative dissolution of their metallic phases, Mo showing higher oxygen affinity than Tc in accordance with their thermodynamic properties. The behavior of Nd and Y gives no evidence of congruent release with the fuel matrix. The concentration of Cs in contacting solutions follows similar trends as Sr and U. Long time experiments indicate that some elements reach saturation with respect to secondary phases. Uranium concentrations seem to be in equilibrium with U(VI) oxohydroxide, excluding when contacting medium contained high concentration of bicarbonate ($10^{-2}$ mol $dm^{-3}$). Np and Pu seem to be solubility limited by the formation of their Ac(IV) hydroxide phases and Tc by the formation of TC(IV) hydroxide phase. Sr is not close to saturation with respect to any secondary phase.

Various authors studied the influence of gamma radiolysis on corrosion of $UO_2$, compared to influence of oxygen or hydrogen peroxide in aqueous environment (Shoesmith et al., 1985; 1989; Sunder et al., 1987). It was found that the values of corrosion potential, at which the stationary state is established, obtained in the presence of gamma radiation, are higher than those reached in solutions containing oxygen or hydrogen peroxide. This shows how significant the contribution of products of water radiolysis to corrosion processes actually is. It was shown that rate of both fuel ($UO_2$) and packaging materials dissolution increases with dose rate (Christensen & Sunder, 2000). Corrosion potential of water - $UO_2$ system increases with dose rate, too. The important parameter is the presence of dissolved oxygen in corrosion system - the corrosion potential increases (among others) faster and to higher values in the presence of dissolved oxygen, when compared to solutions purged with Ar or $N_2O$. As was discussed in section 2.1, in aqueous solutions containing dissolved oxygen, the radical ion $O_2^-$ is formed during radiolysis according to equation (6), which is an effective oxidant of $UO_2$. Thickness of layer of corrosion products on the corroding material surface is directly proportional to the corrosion potential up to the value ~ 0 mV and it does not depend on corrosion rate, nor character of predominating products of products of radiolysis (Christensen & Sunder, 2000). At higher values of corrosion potential, the thickness of surface layer increases due to irradiation, even after reaching the steady state. Oxidation of $UO_2$ proceeds in the presence of oxidative products of gamma radiolysis in two steps:

1. Formation of thin layer with stoichiometry nearing $UO_{2.33}$.
2. Oxidative dissolution of this surface layer, accompanied by formation of soluble forms U(VI) and secondary phases (probably hydrated schoepite) on the surface of electrode. Finally, the steady state of corrosion potential at the value determined by absorbed dose is reached.

Alpha activity of irradiated fuel decreases slower than gamma and beta activities. Therefore, high probability exists that radiation-corrosive processes will proceed for a long time due to alpha radiolysis. Studies of many authors (e.g. Sunder et al., 1997) shown, that rate of corrosion of nuclear fuel and packaging materials due to alpha radiolysis is significantly lower compared to gamma radiolysis. The reason for this is probably the fact that low LET radiation (gamma and beta) leads to higher primary radical yields and lower primary molecular yields than high LET radiation. Therefore, the corrosion effect of alpha radiolysis is caused mainly by hydrogen peroxide, whose oxidative capabilities were observed to be much lower than effects of radicals OH and $O_2^-$. Thus, in many studies, the complicated alpha irradiation is substituted by adding hydrogen peroxide to aqueous solution, because similar behavior of corroding material may be expected (e.g. Daub et al., 2010).

Simulation of the fact that corroding system may be composed of iron (packaging material) and $UO_2$ (fuel) was by authors (Loida et al., 1995) solved by adding powder iron. In this case, the corrosion rate was measured by gas evolution. Authors came to conclusion, that gamma radiolysis contributes to gas evolution much more than alpha radiolysis. They also concluded, that rate of fuel dissolution is in the same order as rate of production of radiolytic gases. Many performed experiments show that corrosion of both packaging materials and fuel is governed by intensity of radiation, degree of burning and degree of non-stoichiometry of corroding material. According to some authors involved in modelling radiolytic processes, it is important to determine dose rates in water layer which is in contact with corroding material, but in practice, such value can be obtained only with great difficulties.

### 3.2 Packaging and barrier materials

Corrosion of fuel and packaging materials is strongly related. Processes leading to degradation of both material types are often studied together (see some remarks on the corrosion of packaging material in the previous section). Opinions of different authors vary, whether the corrosion, even though undoubtedly increased by the presence of radiation field, may actually threaten the stability of packaging/barrier material or not.

#### 3.2.1 Iron and steel

The majority of published works focused on study of radiolytic corrosion of water/iron or steel system use gamma radiation for increased simplicity of experimental setup and high penetration of gamma rays. Usually, doses up to few MGy with dose rates up to tens of kGy have been used. In many experimental studies, pure water as well as synthetic groundwater relevant to anticipated conditions in repository was used. Characterization of corrosion process is often performed via evaluation of changes in corrosion potential, tendency to cracks developing and/or determination of amount and character of corrosion products under various conditions. Generally, the obtained data indicate, that there is rather small effect of radiation on corrosion of stainless steel, especially under deaerated conditions (see fig.1 for example of experimental setup). On the contrary, carbon steel and iron corrode always much faster in radiation field when compared to non-irradiated materials. Similarly to non-radiolytic corrosion, the corrosion induced by radiation strongly depends on temperature of corroding system, as is discussed further in the text.

The effect of gamma radiation on the corrosion of carbon steel and stainless steel contacting high-temperature water has been studied since seventies and it was shown that while irradiation does not seem to have a marked effect on the corrosion of stainless steel, it accelerates the corrosion of carbon steel about 3-4 times (Ershov et al., 1977). It was observed, that the concentration of $H_2O_2$ in irradiated water contacting carbon steel changes according to the fluctuation law with a period of 20-25 h and remains at the same level as for carbon steel. Later, it was concluded that when pure water saturated with air was irradiated at normal temperature by high doses of gamma rays (1-10 MGy) in sealed stainless steel containers, hydrogen and oxygen were formed (Burns et al., 1983). The amounts were less than one tenth of the maximum possible for continuous aqueous radiolysis but the increase in oxygen appearing as gas was less than that equivalent to the hydrogen formed from the water present, indicating that metallic corrosion had occurred. In the absence of radiation no change in gas composition was observed. When the air in solution and in the gas space was replaced by argon or by hydrogen, radiolysis and corrosion were virtually suppressed. When the container was made of mild steel or strips of mild steel were initially introduced into a sealed stainless steel container containing air and water, oxygen was consumed on irradiation, and hydrogen was formed, together with a

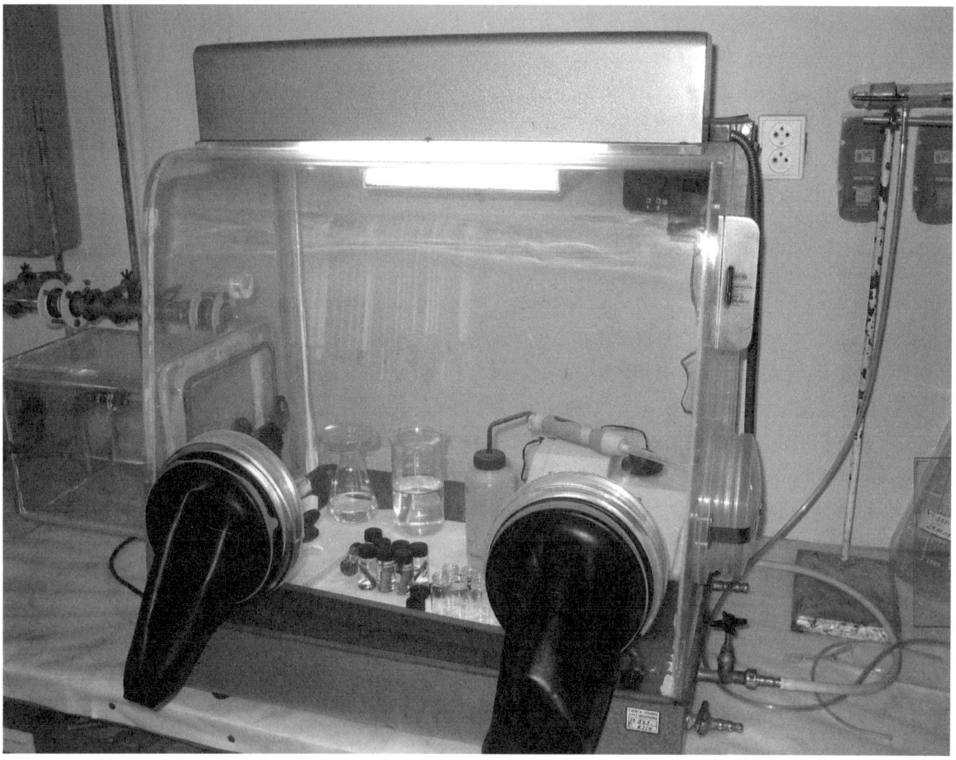

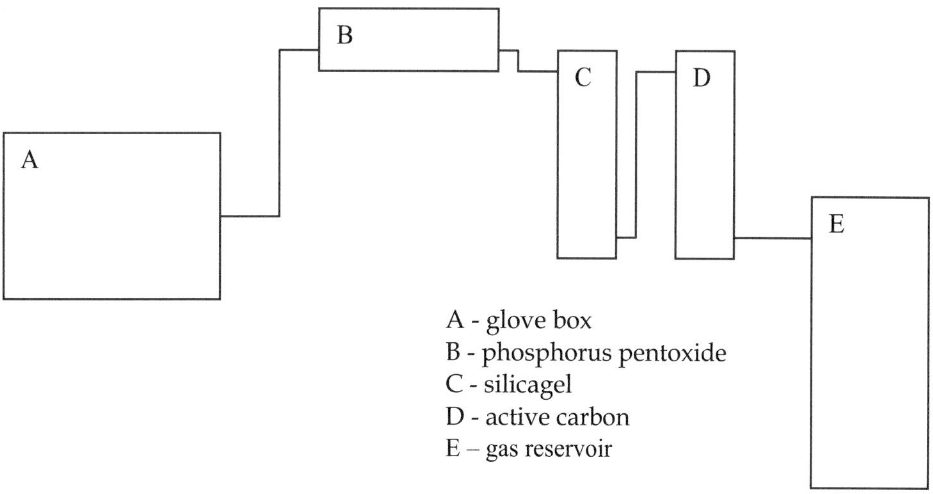


Fig. 1. Experimental setup for working in deaerated conditions. Upper – photo of glove box for working under inert atmosphere with corrosion vials. Lower – example of nitrogen purification prior admitting the gas into glove box

suspended brown oxide, probably $Fe_3O_4$. In the absence of radiation oxygen was consumed and hydrogen was formed, too but both at a lower rate than in the presence of radiation. In this case, unlike the case of stainless steel, the formation of hydrogen was not prevented by replacing the air present with argon. The rate of product formation from either system in the presence of air was found to be proportional to dose-rate, in the applied range 2 - 20 kGy $h^{-1}$ and mass changes of the solids, when measurable, corresponded to the oxygen deficit in the gas phase if the oxide was taken to be $Fe_3O_4$ (Burns et al., 1983).

The corrosion behavior of several iron-base as well as titanium-base alloys was studied in synthetic groundwater at temperatures from 150°C to 250°C and under gamma radiation with dose rates up to 20 kGy $h^{-1}$. The study was performed with regards to repository in basalt (Nelson et al., 1984). Similarly to observations of previously discussed papers, iron-base alloys exhibited a general corrosion rate enhancement factor of 2 to 3 higher when exposed to 250°C synthetic groundwater in the presence of $^{60}Co$ gamma radiation of about 3 kGy $h^{-1}$ than when the radiation was absent. The four different iron-base alloys tested corroded at similar rates. The cast steel containing chromium and molybdenum generally exhibited the greatest resistance to corrosion. Nevertheless, it was concluded that iron-base alloys show rather low corrosion rates under conditions that are much more severe than those anticipated in deep repositories constructed in basalt. Irradiation corrosion studies of titanium in 250°C synthetic groundwater show very low corrosion rates and no tendency to environmentally enhanced cracking. Ahn & Soo (1995) also studied the long term corrosion processes at elevated temperatures. Corrosion properties of carbon steel were evaluated in concentrated synthetic groundwater at 80-150°C, with regards to the use of the steel as a container material in the Yucca Mountain repository project. Corrosion rates were obtained at 150°C under conditions of no irradiation and a gamma irradiation field of 13 kGy $h^{-1}$. Long term experiments with aerated solutions were performed for 4 months. The long-term uniform corrosion rates were 10-40 m $year^{-1}$ under irradiation conditions. Irradiation effects on uniform corrosion were not discernible. Pitting corrosion was also observed but the pitting factor was small. Microstructural effects on corrosion were not significant. During corrosion, hydrogen was generated. Under irradiation conditions, the hydrogen generation was greater and there was an indication that large amounts of hydrogen were absorbed in the steel during the radiation corrosion. Indication of stress corrosion cracking was observed as well as strong evidence for hydrogen embrittlement. Similarly to Nelson et al. (1984), authors concluded that general corrosion rates of the steel appear to be low enough to maintain integrity of containers for an extended period.

Few studies on the mechanisms of corrosion processes on stainless steels in groundwater were performed. Under gamma irradiation, the corrosion potential shifts in the positive direction were observed (Glass et al., 1986). These potential shifts were associated with the radiation-induced production of hydrogen peroxide. Another study of gamma radiation - induced corrosion in deaerated water/carbon steel systems was performed by Neufuss et al. (2006). It was concluded, that kinetics of releasing corrosion products into the water and their sorption on the surface of steel tablets is affected by various factors (redox potential, absorbed dose, temperature, irradiation duration). Concentration of corrosion products in the solution and solid phase was evaluated and corrosion processes were studied in deaerated pure water and synthetic granitic water. Based on results obtained from performed experiments, several general conclusions regarding influence of IR on corrosion of carbon steel/water system were made:

- Under the given conditions, irradiation strongly affects the corrosion kinetics. This effect is clearly observable not only in the corrosion system with dissolved oxygen, but also in the deoxygenated systems.
- Type and distribution of corrosion products depends on some experimental parameters (i.e., temperature, presence of oxygen). Concentration of corrosion products increases with dose of radiation.
- Presence of impurities in granitic water affects corrosion to a small extent.
- Differences in corrosion products concentration were observed between samples saturated with nitrogen and helium - nitrogen is less suitable for radiation experiments under anoxic conditions.

Thorough long term experiments taking several months, aimed to investigate the effect of gamma radiation on the corrosion of carbon steel in repository environments, were carried out (Smart et al., 2008). Corrosion experiments were performed at two temperatures (30°C and 50°C), two dose rates (11 Gy $h^{-1}$ and 300 Gy $h^{-1}$), and in two different artificial groundwaters. In full agreement with previously discussed results, radiation was found to enhance the corrosion rate at both dose rates but the greatest enhancement occurred at the higher dose rate. The corrosion products were predominantly magnetite, with some indications of unidentified higher oxidation state corrosion products being formed at the higher dose rates.

In the work of Daub et al. (2011), the effect of gamma-radiation on the kinetics of carbon steel corrosion has been investigated by characterizing the oxide films formed on steel coupons at 150 °C in aqueous solutions and at two pH values. Though variations in system temperature and pH may affect rates of oxide film growth and dissolution, they do not play a significant role in determining the chemical composition of the oxide film. Gamma irradiation had a major impact on the corrosion process by increasing corrosion potential in the system and therefore determining which oxide phase forms on the steel surface. Specifically at mildly basic pH, the presence of IR leads to the formation of a more passive film composed of a mixture of $Fe_3O_4$ and $\gamma$-$Fe_2O_3$. These results suggest that water radiolysis does not necessarily increase, but rather limits, carbon steel corrosion under the conditions studied. In the similar study (Cuba et al., 2011), the influence of gamma irradiation on the formation of Fe-ions was investigated at different temperatures up to 90°C, with respect to the expected ingress of groundwater into the disposal site with spent nuclear fuel containers. Specifically, the kinetics of $Fe^{2+}$ and $Fe^{3+}$ formation in the presence of IR was studied. It was confirmed, that in the absence of oxygen, or after consumption of oxygen by corrosion in a closed system, corrosion in the presence of $\gamma$-radiation proceeds mainly via reaction with the intermediate products of water radiolysis (mainly $H_2O_2$, and radicals OH, $HO_2$). The radiation increases amount of corrosion products; it affects their composition predominantly via radiation oxidation of ferrous to ferric ions. Characterization of crystalline solid corrosion products revealed that under given conditions, the composition of solid crystalline corrosion products does not depend on temperature, but rather on the presence of other compounds in water. On the other hand, the amount of corrosion products in an irradiated system depends on the temperature and on the dose; other compounds present in water affected the total amount of corrosion products only marginally. In the solid phase formed during contact with deionized water, the predominant crystalline corrosion products were magnetite and lepidocrocite, which is in agreement with (Smart et al., 2008) and partially (Daub et al., 2011). Other published studies discussed here dealing with water-steel system and providing data on various corrosion parameters (e.g. corrosion rate, corrosion potential, hydrogen evolution, etc.) in radiation field generally concluded, that under aerated or deaerated conditions in temperature range 30-250°C, the corrosion is affected by ionizing radiation at dose rates 0.01 kGy. $h^{-1}$ (Smailos, 2002), 0.011-0.3 kGy.$h^{-1}$ (Smart et al., 2008), 3 kGy.$h^{-1}$ (Nelson et al., 1984), 13 kGy.$h^{-1}$ (Ahn and Soo, 1995), which is consistent with observations made by Cuba et al. (2011) at dose rate 0.22 kGy $h^{-1}$ and temperatures 50 and 70°C. Similarly, the slow rates of corrosion processes in irradiated deaerated water at room temperature were also observed by others (Burns et al, 1983; Lapuerta et al., 2005).

### 3.2.2 Other materials

Similarly to corrosion studies of iron and steel, some other materials are also studies for possible use in container manufacture or as matrices for immobilization of irradiated fuel. Studies including gamma radiation predominate, not only due to relative experimental simplicity, but also because among the various types of radiation relevant for nuclear waste disposal, only gamma radiation can penetrate the container wall and affect the corrosion of its outer wall (Michaelis et al., 1998). Besides from already discussed work (Nelson et al., 1984), other authors also studied radiation corrosion or corrosion resistance of various metallic materials considered for construction of engineering barriers. Interesting study of corrosion of Ti and its alloy TiO.ZPd in pH 4.6 aqueous solutions simulating the salt brine environment was performed, because those materials are considered as promising for the manufacture of high level nuclear waste containers which could act as an engineered barrier in a rock salt repository (Michaelis et al., 1998). Despite their extremely high resistance to general corrosion in salt brines, localized corrosion might be critical. Instead of gamma irradiation, authors used in-situ UV-laser illumination with high power density at high local resolution. It was shown that after formation of stable corrosion (amorphous and crystalline oxidic) layers, virtually no further corrosion was observed under given conditions, which confirms older conclusions of (Nelson et al., 1984), i.e. that Ti-based alloys are promising container materials.

For the long-term disposal (immobilization of various radioactive wastes, including actinides, host matrices with a very high resistance to corrosion are needed. Promising matrix material is thorium phosphate-diphosphate (TPD). Irradiation effects of radiation on TPD were studied (Pichot et al., 2001). At low doses (~ kGy), the gamma radiation induced damage (formation of free radicals) could be healed by annealing at higher temperature. At high doses (~ MGy) gamma radiation induced the presence of paramagnetic defects. Effect of accelerated heavy charged particles, used for simulating α-decay effects included modification of the chemical stoichiometry at the surface of the samples, though it was interpreted as a pure ballistic differential sputtering effect involving elastic collisions. The important conclusion was made - due to irradiation, crystalline structure of TPD did not change under given conditions (Pichot et al., 2001).

### 3.3 Nuclear machinery

During operation of nuclear devices, corrosion threat is especially serious. Corrosion may lead to a) decrease of functional properties of materials, and/or to b) significantly increased difficulties in operational control and maintenance due to its contamination by radioactive corrosion products. To fully explore possible consequences of radiolytic corrosion in nuclear/irradiation devices, irradiation with heavy particles must be employed aside from gamma radiation. For example, spallation neutron sources generate a mixed radiation environment when a beam of high energy particles (e.g. protons or deuterons) hits a heavy metal target. Radiolysis results when these primary and secondary (spallation) particles lose energy by Coulomb interaction with the electrons in the hydrogen and oxygen atoms of water (Lillard, 2003). In either case, the corrosion threatens reliability and safeguard of operation of nuclear device.

Corrosion of spallation neutron sources was studied (Lillard et al., 2000). In those sources, high energy neutrons are produces, via interaction of accelerated protons with convenient target material. Cooling loop, which keeps low temperature of target material is usually constructed from stainless steel and filled with deionized water, serving also as moderator of neutrons. However, radiolysis of coolant water may result in increase in corrosion of construction materials. Referred work is interesting also from experimental point of view: authors irradiated water and let it react with corrosion sonds, which were placed outside of radiation field and shielded. By this, authors aimed at eliminating of direct radiation influence on selected corrosive materials - stainless steel and alloys of aluminium, tungsten, nickel and copper. The effects of water system fabrication materials, hydrogen water chemistry, and pre-cleaning of the water system on radiation corrosion rate have been demonstrated to be dramatic. Study of corrosion rates in real time resulted in observation, that under given conditions, the most significant corrosion occurs at materials containing copper and aside from radiation, the corrosion rate is affected mainly by amount of impurities present in irradiated water. Other findings indicate that proton irradiation and the resulting water radiolysis products do not influence passive film formation and reduction (Lillard, 2003).

The effect of IR on carbon steel corrosion as an important materials issue in nuclear reactors was investigated (Daub et al., 2010) at pH 10.6. The effect of gamma-radiation on corrosion kinetics was compared with that of chemically added $H_2O_2$, which is considered to be the key radiolytically produced oxidant at room temperature (see section 2.1 for details on its formation). It was confirmed that $H_2O_2$ is really the key radiolysis product controlling carbon steel corrosion. The discussed study illustrates that the corrosion rate of carbon steel in a gamma-radiation environment at alkaline pH and room temperature can be predicted from the dependence of corrosion potential and polarization resistance on $H_2O_2$ concentration, if the concentration of radiolytically produced $H_2O_2$ can be determined.

## 4. Conclusions

It was demonstrated, that corrosion processes in aqueous environment may be strongly affected by the presence of radiation field. IR contributes to corrosion mainly via production of highly reactive intermediates of water radiolysis. Similarly to chemical corrosion, the effects of radiation induced corrosion are also enhanced when the corroding system operates at elevated temperatures and they depend on pH value of a liquid irradiated system. The effects of radiation corrosion are also strongly dependent on dose, dose rate and LET of applied radiation. Due to high (long term) level of corrosion caused by radiation, it is important aspect to consider in all steps of nuclear fuel cycle, especially in storage / disposal of irradiated nuclear fuel. Radiation corrosion may also present important technological problem in operation of nuclear reactors and various irradiation devices.

## 5. Acknowledgements

Authors gratefully acknowledge financial support of Ministry of Education, Youth and Sports of the CR, project MSM 68-4077-0040.

## 6. References

Ahn, T.M. & Soo, P. (1995). Corrosion of low-carbon cast steel in concentrated synthetic groundwater at 80 to 150°C. *Waste Manage* 15, pp. 471-476, ISSN 0956-053X

Allen, A.O. (1961). *The Radiation Chemistry of water and aqueous solutions.* Van Nostrand, ISBN 0442003013, Princeton

Anbar, M. & Thomas, J.K. (1964). Pulse radiolysis studies of aqueous sodium chloride solutions. *J. Phys. Chem.* 68, pp. 3829-3835, ISSN 0022-3654

Andersson, J. (1999). *Data and data uncertainties.* SKB technical report TR-99-09, Stockholm, Sweden

Belloni, J.; Mouard, H.; Gobert, F.; Larbre, J.P.; Danarque, A.; De Waele, V.; Lampre, I.; Mariguier, J.L.; Mostafavi, M.; Bourdon, J.C.; Bernard, M.; Borie, H.; Garvey, T.; Jacquemard, B.; Leblond, B.; Lepercq, P.; Omeich, M.; Roch, M.; Rodier, J. & Roux, R. (2005). ELYSE-A picosecond electron accelerator for pulse radiolysis research. *Nucl. Instr. Methods in Phys. Res. A* 539, pp. 527-539, ISSN 0168-9002

Bruno, J.; Cera, E.; Grive, M.; Eklund, U-B. & Eriksen, T. (1999). *Experimental determination and chemical modelling of radiolytic processes at the spent nuclear fuel/water interface.* SKB technical report TR-99-26, Stockholm, Sweden

Burns, W.G.; Marsh, W.R. & Walters, W.S. (1983). The ϒ irradiation-enhanced corrosion of stainless and mild steels by water in the presence of air, argon and hydrogen. *Radiat. Phys. Chem.* 21, pp. 259 -279, ISSN 0969-806X

Buxton, G.V. (1972). Nanosecond pulse radiolysis of aqueous solutions containing proton and hydroxil radical scavengers. *Proc. R. Soc. Lond. A.* 328, pp. 9-21; In: Farhatazis & Rodgers, M.A.J. (Eds.) (1987). *Radiation Chemistry. Principles and Applications.* VCH Publishers, ISBN 0895731274, New York

Buxton, G.V.; Greenstock, C.L.; Helman, W.P. & Ross, A.B. (1988). Critical review of rate constants for reactions of hydrated electrons, hydrogen atoms and hydroxyl radicals in aqueous solution. *J. Phys. Chem. Ref. Data* 17, pp. 513-886, ISSN 0047-2689

Crowell, R.A.; Shkrob, I.A.; Oulianov, D.A.; Korovyanko, O.; Goszatola, D.J.; Li, Y. & Rey-de-Castro, R. (2005). Motivation and development of ultrafast laser-based accelerator techniques for chemical physics research. *Nucl. Instr. Meth. in Phys. Res. B* 241, pp. 9-13, ISSN 0168-583X

Christensen H. & Sunder, S. (2000). Current state of knowledge of water radioloysis effects on spent nuclear fuel corrosion. *Nucl. Technol.* 131, pp. 102-119, ISSN 0029-5450

Cuba, V.; Silber, R.; Mucka, V.; Pospisil, M.; Neufuss, S.; Barta, J. & Vokal, A. (2011). Radiolytic formation of ferrous and ferric ions in carbon steel - deaerated water system. *Radiat. Phys. Chem.* 80, pp. 440-445, ISSN 0969-806X

Daub, K.; Zhang, X.; Noel, J.J. & Wren, J.C. (2010). Effects of ϒ- radiation versus $H_2O_2$ on carbon steel corrosion. *Electrochim. Acta* 55, pp. 2767 -2776, ISSN 0013-4686

Daub, K.; Zhang, X.; Noel, J.J. & Wren, J.C. (2011). Gamma-radiation-induced corrosion of carbon steel in neutral and mildly basic water at 150 °C. *Corros. Sci.* 53, pp. 11-16, ISSN 0010-938X

Domae, M.; Katsumura, Y.; Ishiguere, K. & Byakov, V.M. (1996). Modelling of primary chemical processes of water radiolysis and simulation by spur diffusion model. *Radiat. Phys. Chem.* 48, pp. 487-495, ISSN 0969-806X

Domski, P.S. (2003). *In-Package Chemistry Abstraction.* OCRWM Analysis, ANL-EBS-MD-000037

Draganic, I.G. & Draganic, Z.D. (1971). *The radiation chemistry of water.* Academic Press Inc., ISBN 0122216504, New York.

Erskov, B.G. & Gordeev, A.V. (1959). A model for radiolysis of water and aqueous solutions of $H_2$, $H_2O_2$, and $O_2$. *Radiat. Phys. Chem.* 77, pp. 928-935, ISSN 0969-806X

Erskov, B.G.; Milaev, A.I.; Petrosyan, V.G.; Kartashov, N.I.; Glasunov, P.Ya. & Tevlin, S.A. (1985). The effect of radiation on corrosion of steel in high-temperature water. *Radiat. Phys. Chem.* 26, pp. 587-590, ISSN 0969-806X

Federal Ministry of Economics and Technology (2008). *Final disposal of high-level radioactive waste in Germany- the Gorleben repository project.* Edit. Federal Ministry of Economics and Technology (BMWi), pp. 8-62, Berlin

Ferradini, Ch. & Jay-Gerin, J.P. (2000). The effect of pH on water radiolysis: a still open question-a minireview. *Res. Chem. Intermed.* 26, pp. 549-565, ISSN 0922-6168

Glass, R.S.; Overturf, G.E.; Van Konynenburg, R.A. & McCright, R.D. (1986). Gamma radiation effects on corrosion-I. electrochemical mechanisms for the aqueous corrosion processes of austenitic stainless steel relevant to nuclear waste disposal in TUFF. *Corros. Sci.* 26, pp. 577-590, ISSN 0010-938X

Haarlamert, T. & Zacharias H. (2009). Application of high harmonic radiation in surface science. *Current Opinion in Solid State and Mater. Sci.* 13, pp. 13-27, ISSN 1359-0286

Huie, R.E. & Clifton, C.L. (1990). *Temperature dependence of the rate constants for reactions of the sulfate radical, $SO_4^-$, with anions.* J. Phys. Chem. 94, 8561 - 8567, ISSN 0022-3654

Jiang, P.Y.; Katsumura, Y.; Nagaishi, R.; Domae, M.; Ishikawa, K. & Ishigure, K. (1992). Pulse radiolysis study of concentrated sulfuric acid solutions. *J. Chem. Soc. Faraday Trans.* 88, pp. 1653-1568, ISSN 1463-9076

Kaesche, H. (1985). *Metallic Corrosion,* NACE, ISBN 0915567121, 2nd ed. Houston

Katsumura, Y. (2001). Radiation chemistry of concentrated inorganic aqueous solutions. In: *Radiation chemistry, present status and future trends,* Jonah, C.D. & Rao, B.S.M. (eds), Elsevier, ISBN 978-0444829023, Amsterdam

LaVerne, J.A. (2000). OH radicals and oxidizing products in the gamma radiolysis of water. *Radiat. Res.* 153, pp. 196-200, ISSN 0033-7587

Liu, J. & Neretnieks, I. (1996). A model for radiation energy deposition in natural uranium-bearing systems and its consequences to water radiolysis. *J. Nucl. Mater.* 231, pp. 103-112, ISSN 0022-3115

Liu, X.; Zhang, G. & Thomas, J.K. (1997). *Spectroscopic studies of electron and hole trapping in zeolites: formation of hydrated electrons and hydroxil radicals.* J. Phys. Chem. B 101, 2182-2194, ISSN 1520-6106

Lillard, R.S., (2003). Influence of water radiolysis products on passive film formation and reduction in a mixed radiation environment. *Corros. Eng. Sci. Techn.* 38, pp. 192-196, ISSN 1743-2782

Lillard, R.S.; Pile, D.L. & Butt, D.P. (2000). The corrosion of materials in water irradiated by 800 MeV protons. *J. Nucl. Mater.* 278, 277-289, ISSN 0022-3115

Loida, A.; Grambow, H.; Beckeis, H. & Dressler, P. (1995). Processes controlling radionuclide release from spent fuel. Scientific basis for nuclear waste management XVIII, Murakami, T.; Ewing, R.E. (Eds), *Mater. Res. Soc. Symp. Proc.* 353, pp. 577-584, ISSN 0272-9172

McElroy, W.J. & Waygood, S.J. (1990). Kinetics of the reactions of the $SO_4^-$ radical with $SO_4$, $S_2O_8^{2-}$, $H_2O$ and $Fe^{2+}$. *J. Chem. Soc. Faraday Trans.* 86, pp. 2557 - 2564, ISSN 1463-9076

Michaelis, A.; Kudelka, S. & Schultze, J.W. (1998). Effect of y-radiation on the passive layers of Ti and TiO.2Pd container-materials for high-level waste disposal. *Electrochim. Acta* 43, pp. 119-130, ISSN 0013-4686

Motl, A. (2004). *Introduction to radiation chemistry.* CTU publishing, ISBN 8001029298, Prague (*in Czech*)

Nelson, J.L; Westerman, R.E. & Gerber, F.S. (1984). Irradiaton-corrosion evaluation of metals for nuclear waste package applications in Grande Ronde Basalt groundwater. *Mater. Res. Soc. Symp. Proc.* 26, pp. 121-128, ISSN 0272-9172

Neufuss, S.; Cuba, V.; Silber, R.; Mucka, V.; Pospisil, M. & Vokal, A. (2006). Experimental simulation of possible radiation-corrosive processes in container with spent nuclear fuel after groundwater ingress. *Czech. J. Phys.* 56, pp. D365-D372, ISSN 0011-4626

Ollila, K. (1992). SIMFUEL dissolution studies in granitic groundwater. *J. Nucl. Mater* 190, pp. 70-77, ISSN 0022-3115

Pan, Y.-M.; Brossia, C.S.; Cragnolino, G.A.; Dunn, D.S.; Jain, V. & Sridhar, N. (2002). Evolution of solution chemistry through interactions with waste package internal structural components. *Mat. Res. Soc. Symp. Proc.* 713, pp. 121-127, ISSN 0272-9172

Parks, G.A. & Pohl, D.C. (1988). Hydrothermal solubility of uranite. *Geochim. Cosmochim. Acta*, 52, pp. 863-875, ISSN 0016-7037

Pichot, E.; Dacheux, N.; Emery, J.; Chaumont, J.; Brandel, V. & Genet, M. (2001). Preliminary study of irradiation effects on thorium phosphate-diphosphate. *J. Nucl. Mater.* 289, pp. 219-226, ISSN 0022-3115

Pikaev, A.K. (1986). *Contemporary radiation chemistry: radiolysis of gases and liquids.* Nauka, Moscow (*in Russian*)

Pikaev, A.K.; Kabakchi, S.A. & Egorov, G.F. (1988). Some radiation chemical aspects of nuclear engineering. *Radiat. Phys. Chem.* 31, pp. 789-803, ISSN 0969-806X

Pitter, P. (1999). *Hydrochemistry*, ICT Press, ISBN 978-80-7080-701-9, 3rd ed., Prague (*in Czech*)

Sawasaki, T.; Tanabe, T.; Yoshida, T. & Ishida., R. (2003). Application of gamma radiolysis of water for $H_2$ production. *J. Radioanal. Nucl. Chem.* 255, pp. 271-274, ISSN 0236-5731

Sehested, K. & Christensen, H. (1987). *The radiation chemistry of water and aqueous solutions at elevated temperatures.* Fielden E.M. (Ed.) Proc. 8th Int. Congr. of Radiation Research, ISBN 0-85066-399-7-vol. 1 and ISBN:0-85066-385-7-vol. 2, Edinburg, pp. 199-204

Shoesmith, D.W.; Kolar, M.; King, F. (2003). A mixed-potential model to predict fuel (uranium dioxide) corrosion within a failed nuclear waste container. *Corrosion 59,* pp. 802-816, ISSN 0010-9312

Shoesmith, D.W.; Sunder, S.; Bailey, M.G. & Wallace, G.J. (1989). The corrosion of nuclear fuel ($UO_2$) in oxygenated solutions. *Corros. Sci.* 29, pp. 1115-1128, ISSN 0010-938X

Shoesmith, D.W.; Sunder, S.; Johnson, J.H. & Bailey, M.G. (1985). Oxidation of CANDU $UO_2$ fuel by alpha-radiolysis products of water. Scientific basis for waste management IX, Werme, L.O. (Ed), *Mater. Res. Soc. Symp. Proc.* 50, pp. 309-316, ISSN 0272-9172

Sieferman, K.R.; Liu, Y.; Lugovoy, E.; Link, O.; Fazebel, M.; Buck, U.; Winter, B. & Abel, B. (2010). Binding energies, lifetimes and implications of bulk and interface solvated electrons in water. *Nature Chem.* 2, pp. 274-279, ISSN 1755-4330

Smailos, E. (2002). Influence of gamma radiation on the corrosion of carbon steel, heat-generating nuclear waste packaging in salt brines. In: *Effects of radiation and environmental factors on the durability of materials in spent fuel storage and disposal,* IAEA-TECDOC-1316, pp. 131 -140, ISBN 92–0–113802–4, IAEA, Wienna

Smart, N.R.; Rance, A.P. & Werme, L.O. (2008). The effect of radiation on the anaerobic corrosion of steel. *J. Nucl. Mater.* 379, pp. 97 - 104, ISSN 0022-3115

Spinks, J.W.T. & Woods, R.J. (1990). *Introduction to radiation chemistry.* Wiley Interscience, ISBN 978-0471614036, 3rd ed, New York

Stepanov, S.V. & Byakov, V.M. (2005). On the mechanism of formation of intratrack yields of water radiolysis products upon irradiation with fast electrons and positrons: 1. model formulation. *High Energ. Chem.* 39, pp. 131-136, ISSN 0018-1439

Sunder, S. & Christensen, H. (1993). Gamma radiolysis of water solutgions relevant to the nuclear fuel waste management program. *Nucl. Tech.* 104, pp. 403-413, ISSN 0029-5450

Sunder, S.; Shoesmith, D.W.; Christensen, H. & Miller, N.H. (1992). Oxidation of $UO_2$ fuel by the products of gamma radiolysis of water. *J. Nucl. Mater.* 190, 78-86, ISSN 0022-3115

Sunder, S.; Shoesmith, D.W.; Johnson, J.H.; Bailey, M.G.; Wallace, G.J. & Snaglewski, A.P. (1987). Oxidation of CANDU™ by the products of alpha radiolysis of groundwater. Scientific basis for waste management X, Bates, J.K. & Seefeldt, W.B. (Eds), *Mater. Res. Soc. Symp. Proc.* 84, pp. 103-113, ISSN 0272-9172

Vladimirova, M.V., (1990). Radiation Chemistry of Actinides. *J. Radioanal. Nucl. Chem.* 143, pp. 445-454, ISSN 0236-5731

Watanabe, R. & Saito, K. (2001). Monte Carlo simulation of water radiolysis in oxygenated condition for monoenergetic electrons from 100 eV to 1 MeV. *Radiat. Phys. Chem.* 62, pp. 217-228, ISSN 0969-806X

Wishart, J.F. & Nocera, D.G. (Eds.) (1998). *Photochemistry and radiation chemistry (complementary methods for the study of electron transfer).* Advances in Chemistry series 254, American Chemical Society, ISBN 0-8412-3499-X, Washington

Wishart, J.F. & Rao, B.S.M. (Eds.) (2010). *Recent trends in radiation chemistry.* World Scientific Publishing Co. Pte. Ltd., ISBN 10-981-4282-07-3, Singapore

Woods, R.J. & Pikaev, A.K. (1994). *Applied radiation chemistry: radiation processing*. John Wiley and Sons, Inc., ISBN 0-471-54452-3, New York

Yang, J., Kondoh, T., Kan, K. & Yoshida, Y. (2011). Ultrafast pulse radiolysis. *Nucl. Instr. Meth. in Phys. Res. A* 629, pp. 6-10, ISSN 0168-9002

Yang, J.; Kondoh, T.; Norfizava, K.; Yoshida, Y. & Tagawa, S. (2009). Breaking time-resolution limits in pulse radioloysis. *Radiat. Phys. Chem.* 78, pp. 1164-1168, ISSN 0969-806X

# 3

# Numerical Analysis of Melting/Solidification Phenomena Using a Moving Boundary Analysis Method X-FEM

Akihiro Uchibori and Hiroyuki Ohshima
*Japan Atomic Energy Agency*
*Japan*

## 1. Introduction

Melting/solidification or dissolution/precipitation phenomena appear in several candidate technologies for fast reactor fuel cycle. The fuel cycle consists of reprocessing of spent fuels and fuel fabrication. Combination of the advanced aqueous reprocessing system and the simplified pelletizing fuel fabrication system is a most promising concept for recycling mixed oxide fuels (Sagayama, 2007; Funasaka & Itoh, 2007). The advanced aqueous reprocessing includes the process of dissolution of sheared spent fuels in a nitric acid. The most part of uranium is recovered from the dissolved solution by crystallization. There are dissolution/precipitation phenomena also in the metal electro-refining pyrochemical reprocessing, which is a candidate technology for reprocessing of metal fuels (Kofuji, 2010). In this technology, uranium and plutonium are recovered from spent metal fuels by electrolysis. A metal fuel slug is formed by the injection casting method. In this method, an injected molten fuel alloy solidifies in the mold. Melting/solidification phenomena appear also in the process of high-level radioactive wastes disposal (Kozaka & Tominaga, 2005). High-level radioactive wastes are fed into a melter and mixed with a glass melt. The molten mixture is subsequently cooled in a canister. Melting/solidification or dissolution/precipitation phenomena have been applied not only in the above-mentioned technologies but also in many engineering fields.

Numerical analysis considering melting/solidification or dissolution/precipitation is very useful to evaluate operating conditions and designs of equipment configuration in these technologies. However, there is a difficulty that a moving solid-liquid interface exists in such phase change problems. Reproducing move of the solid-liquid interface in numerical analysis is a challenging problem. The existing numerical analysis methods for melting/solidification problems can be classified into two groups: a moving mesh method and an enthalpy method. In the former group, an analysis mesh is reconstructed during calculation so that nodes overlap with the solid-liquid interface. Independent energy equations of the each phase are coupled by an appropriate boundary condition at the solid-liquid interface. Lynch & O'Neill (1981) developed a moving mesh finite element method, which is a typical method of the former group. Sampath & Zabaras (1999) applied the moving mesh finite element method to a three-dimensional directional solidification problem. A variable space network method developed by Murray & Landis (1959) and a front tracking ALE (Arbitrary Lagrangian Eulerian) method developed by Jaeger & Carin (1994) are also classified as the former group. While the moving mesh methods give high accuracy in predicting an interface position, there is the drawback that the algorithm of re-meshing is complicated especially in the case of multi-dimensional problems. When a shape of the solid-liquid interface becomes complex, we have to pay attention to deformation of elements. Therefore, the moving mesh methods are frequently used for directional melting/solidification problems. The latter group (i.e. an enthalpy method) introduces an artificial heat capacity containing latent heat for phase change. This enables us to eliminate a boundary condition at the solid-liquid interface. The enthalpy method requires only a single energy equation. Numerical analysis using the enthalpy method may be found in the paper by Comini et al. (1974) or by Rolph & Bathe (1982). The enthalpy method is quite popular for multi-dimensional problems because re-meshing is not required. However, there is also the drawback that isothermal phase change phenomena cannot be modeled consistently. This is due to the inevitable assumption that phase change occurs within a certain range of temperature. Application of the above-mentioned two types of the methods is limited by their inherent drawbacks.

In application to the actual problems of the nuclear fuel cycle, a numerical analysis method needs to be applicable to multi-dimensional problems which involve complex move of the solid-liquid interface. A numerical analysis method using a fixed mesh can be simply applied to multi-dimensional problems even if they involves complex move of the interface. However, existence of a discontinuous temperature gradient at the solid-liquid interface complicates calculation of heat conduction and interface tracking in a fixed mesh. The eXtended Finite Element Method (X-FEM), which is a fixed-mesh-based-method, can overcome this difficulty. This method introduces an enriched finite element interpolation to represent the discontinuous temperature gradient in the element. The enriched finite element interpolation consists of a standard shape function and a signed distance function. This makes it possible to track the moving solid-liquid interface accurately without re-meshing. The X-FEM has the advantages of both the moving mesh method and the enthalpy method. Moës et al. (1999) developed the X-FEM for arbitrary crack growth problems. Merle & Dolbow (2002) and Chessa et al. (2002) applied the X-FEM to melting/solidification problems. Further, Chessa & Belytschko (2003) simulated a two-phase flow problem involving bubble deformation successfully by using the X-FEM. These researches indicate that the X-FEM is widely applicable to moving interface problems.

The X-FEM is most effective as a tool to evaluate the melting/solidification or the dissolution/precipitation processes appearing in the nuclear fuel cycle. But the way to apply the X-FEM to phase change problems has not been fully elucidated. In this study, application of the X-FEM to melting/solidification problems was discussed. Formulation of the enriched finite element interpolation and construction of the finite element equation using it are reported below. The technique of quadrature and the method to solve the problems involving liquid flows were constructed in the present work. The numerical solutions of the example problems, i.e. a one-dimensional Stefan problem and solidification in a two-dimensional square corner, were compared to the existing solutions to verify validity of the proposed numerical method. Verification to the melting/solidification problem involving natural convection was also conducted.

## 2. Numerical methods

### 2.1 Governing equations

We consider a melting/solidification problem shown in Fig. 1. The domain $\Omega$ is divided into the solid region $\Omega_s$ and the liquid region $\Omega_l$. The boundary of the domain $\Omega$ and the interface between the solid region and the liquid region are denoted by $\Gamma$ and $\Gamma_I$, respectively. The governing equations for this problem are as follows:

$$\nabla \cdot \mathbf{u} = 0 \tag{1}$$

$$\rho\left(\frac{\partial \mathbf{u}}{\partial t} + \mathbf{u} \cdot \nabla \mathbf{u}\right) = -\nabla p + \mu \nabla^2 \mathbf{u} + \rho \mathbf{f} \tag{2}$$

$$\rho c_p \left(\frac{\partial T}{\partial t} + \mathbf{u} \cdot \nabla T\right) = \lambda \nabla^2 T \tag{3}$$

where $\mathbf{u}$ is the velocity vector, $\rho$ is the density, $t$ is the time, $p$ is the pressure, $\mu$ is the viscosity, $\mathbf{f}$ is the external force, $c_p$ is the specific heat, $T$ is the temperature and $\lambda$ is the thermal conductivity. Equations (1) and (2) are applied only to the liquid phase.

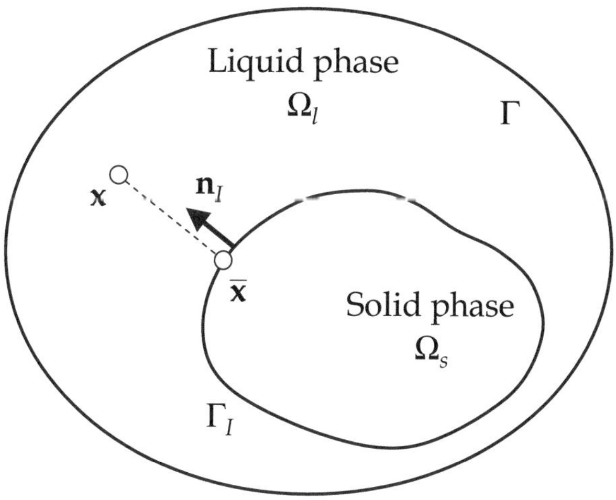


Fig. 1. Melting/solidification problem

A heat balance condition at the solid-liquid interface is written as

$$\rho_s L V_I = \lambda_s \left[\frac{\partial T}{\partial n}\right]_{Is} - \lambda_l \left[\frac{\partial T}{\partial n}\right]_{Il} \tag{4}$$

where $L$ is the latent heat for phase change, $V_I$ is the interface velocity (normal direction to the interface) and $n$ is the normal direction to the solid-liquid interface. Subscript $s$ stands for the solid, $l$ the liquid and $I$ the solid-liquid interface. Equation (4) means that the net heat flux at the solid-liquid interface is translated into the latent heat for phase change per unit time (see Fig. 2). The interface velocity can be obtained from Eq. (4). A temperature condition at the solid-liquid interface is given by

$$T = T_m \tag{5}$$

where $T_m$ is the melting temperature. Equation (5) is used as a constraint condition to Eq. (3), as described later. Melting/solidification problems appearing in the nuclear fuel area are basically described by Eq. (1) to (5).

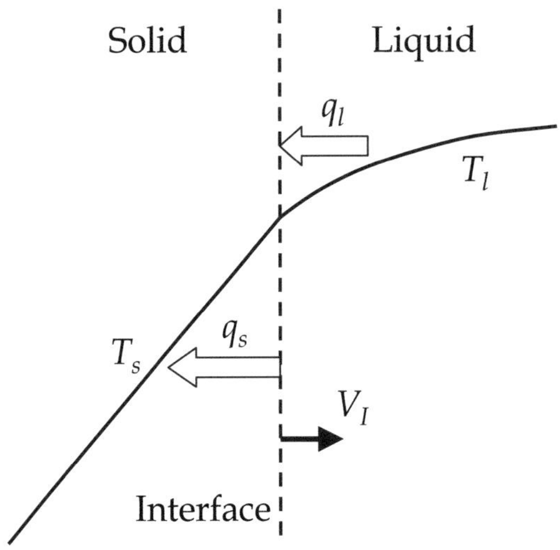


Fig. 2. Temperature profile and heat flux near solid-liquid interface

### 2.2 Enriched finite element interpolation

In melting/solidification problems, the temperature gradient is discontinuous at the solid-liquid interface (see Fig. 2). A basic idea of the X-FEM is to introduce an enriched finite element interpolation to represent the discontinuous gradient of the temperature. In this section, we report the formulation described in the paper by Chessa et al. (2002) and present the temperature profile represented by the enriched finite element interpolation in the case of a two-node line element.

We define the element crossed by the solid-liquid interface as a fully enriched element. The node of the fully enriched element is defined as an enriched node. Figure 3 shows definition of the enriched element and the enriched node in the case of a one- or two-dimensional mesh. The enriched finite element interpolation is based on a standard interpolation:

$$T(\mathbf{x},t) = \sum_{i=1}^{n} N_i(\mathbf{x}) T_i(t) \tag{6}$$

where $i$ is the node number, $n$ is the number of nodes, $N_i(\mathbf{x})$ is the shape function and $T_i(t)$ is the nodal value of the temperature:

$$T_i(t) = T(\mathbf{x}_i, t) \tag{7}$$

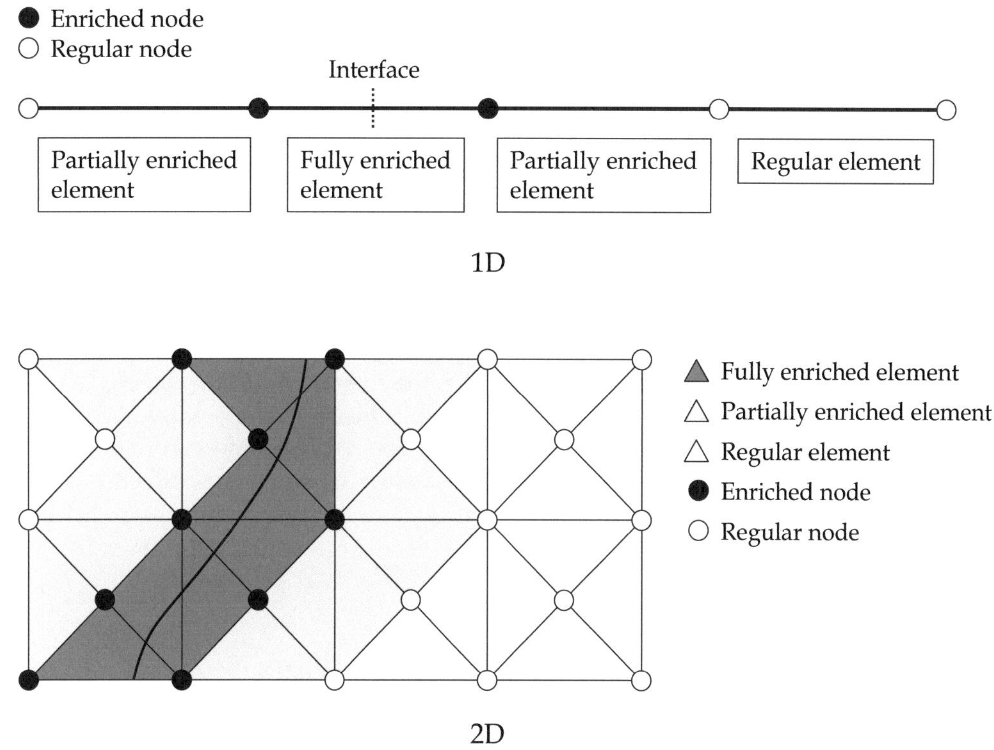


Fig. 3. Definition of enriched elements and nodes

Figure 4(a) shows the temperature profile interpolated by Eq. (6) in the case of a two-node line element. To construct a discontinuous interpolation, we introduce a signed distance function $\phi(\mathbf{x},t)$:

$$\phi(\mathbf{x},t) = \min_{\bar{\mathbf{x}} \in \Gamma_I} \|\mathbf{x}-\bar{\mathbf{x}}\| \operatorname{sign}\left[\mathbf{n}_I \cdot (\mathbf{x}-\bar{\mathbf{x}})\right] \tag{8}$$

where $\| \ \|$ is the Euclidian norm, $\bar{\mathbf{x}}$ is the coordinate on the solid-liquid interface and $\mathbf{n}_I$ is the outward normal unit vector from the solid region (see Fig. 1). This definition means the shortest distance from $\mathbf{x}$ to the solid-liquid interface. The signed distance function varies as follows:

$$\phi(\mathbf{x},t) \begin{cases} <0 & \mathbf{x} \in \Omega_s \\ =0 & \mathbf{x} \in \Gamma_I \\ >0 & \mathbf{x} \in \Omega_l \end{cases} \tag{9}$$

Here, we define an enrichment function $\psi_i(\mathbf{x},t)$:

$$\psi_i(\mathbf{x},t) = N_i(\mathbf{x})\left(|\phi(\mathbf{x},t)| - |\phi(\mathbf{x}_i,t)|\right) \tag{10}$$

Figure 4(b) shows the profile of the enrichment function $\psi_1$ and $\psi_2$. As can be seen, the enrichment function is discontinuous at the solid-liquid interface. Figure 4(c) shows the profile of $\psi_1 + \psi_2$. This summation becomes linear between the interface and the node. Adding the enrichment function to the standard interpolation function, we obtain the enriched finite element interpolation:

$$T(\mathbf{x},t)=\sum_{i=1}^{n}N_i(\mathbf{x})T(\mathbf{x}_i,t)+\sum_{j=1}^{n_e}\psi_j(\mathbf{x},t)a_j(t) \tag{11}$$

where $n_e$ is the number of enriched nodes and $a_j(t)$ is the additional degrees of freedom. The temperature profile interpolated by Eq. (11) is shown in Fig. 4(d). While the two-node regular element has two degrees of freedom (i.e. the unknowns, $T_1$ and $T_2$), the fully enriched element has four degrees of freedom (i.e. the unknowns, $T_1$, $T_2$, $a_1$ and $a_2$). The additional degrees of freedom, $a_1$ and $a_2$, are decided so that the temperature at the solid-liquid interface becomes the melting temperature. The way to constrain the condition of the temperature at the solid-liquid interface to the finite element equation is described in section 2.3.2.. The gradient of the temperature is written as

$$\begin{aligned}\nabla T(\mathbf{x},t)=&\sum_{i=1}^{n}\nabla N_i(\mathbf{x})T(\mathbf{x}_i,t)\\&+\sum_{j=1}^{n_e}a_j(t)\Big[\nabla N_j(\mathbf{x})\big(|\phi(\mathbf{x},t)|-|\phi(\mathbf{x}_j,t)|\big)+N_j(\mathbf{x})\nabla\big(|\phi(\mathbf{x},t)|-|\phi(\mathbf{x}_j,t)|\big)\end{aligned} \tag{12}$$

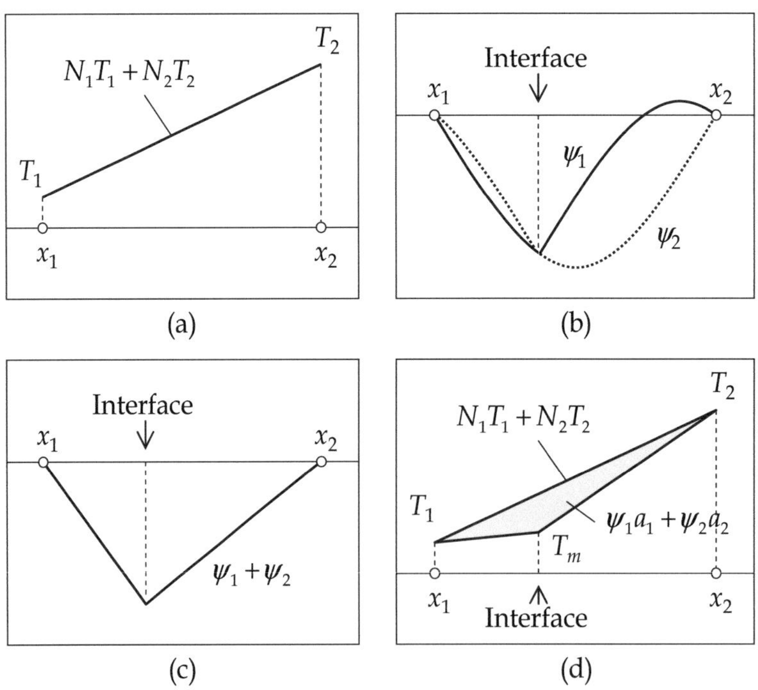


Fig. 4. Enriched finite element interpolation in two-node line element

Not all of the nodes are enriched in the element adjacent to the fully enriched element. This partially enriched element is also shown in Fig. 3. If we apply the enriched interpolation defined as Eq. (11) to the partially enriched element, the unnecessary numerical value is added to the desired profile. This problem has been discussed in the paper by Chessa et al. (2003). We have used their method to avoid the incorrect interpolation.

### 2.3 Finite element equation

#### 2.3.1 Matrix form

Based on the enriched finite element interpolation, i.e. Eq. (11), a finite element equation of the energy equation is derived here. For describing the finite element equation in matrix form, it is convenient to rewrite Eq. (11) as

$$T(\mathbf{x},t)=\bar{\mathbf{N}}(\mathbf{x},t)\bar{\mathbf{T}}(t) \tag{13}$$

where

$$\bar{\mathbf{N}}(\mathbf{x},t)=\begin{bmatrix} N_1 & N_2 & \cdots & N_n & \psi_1 & \psi_2 & \cdots & \psi_{n_e} \end{bmatrix} \tag{14}$$

$$\bar{\mathbf{T}}(t)=\begin{bmatrix} T_1 & T_2 & \cdots & T_n & a_1 & a_2 & \cdots & a_{n_e} \end{bmatrix}^T \tag{15}$$

The matrix form of Eq. (3) is

$$\left(\frac{1}{\Delta t}\mathbf{M}^{n+1}+\mathbf{C}^{n+1}+\mathbf{K}^{n+1}\right)\bar{\mathbf{T}}^{n+1}=\frac{1}{\Delta t}\mathbf{M}^{n}\bar{\mathbf{T}}^{n} \tag{16}$$

where $\mathbf{M}$ is the mass matrix, $\mathbf{C}$ is the coefficient matrix of the convection term and $\mathbf{K}$ is the coefficient matrix of the heat conduction term. These coefficient matrices consist of the shape function and the enrichment function. The mass matrix in the left hand side is written as

$$\mathbf{M}^{n+1}=\begin{bmatrix} \mathbf{M}_1^{n+1} & \mathbf{M}_2^{n+1} \\ \mathbf{M}_3^{n+1} & \mathbf{M}_4^{n+1} \end{bmatrix} \tag{17}$$

where

$$\begin{aligned}
\left[\mathbf{M}_1^{n+1}\right]_{ij} &= \int_\Omega \rho c_p N_i N_j d\Omega \\
\left[\mathbf{M}_2^{n+1}\right]_{ij'} &= \int_\Omega \rho c_p N_i \psi_{j'}^{n+1} d\Omega \\
\left[\mathbf{M}_3^{n+1}\right]_{i'j} &= \int_\Omega \rho c_p \psi_{i'}^{n+1} N_j d\Omega \\
\left[\mathbf{M}_4^{n+1}\right]_{i'j'} &= \int_\Omega \rho c_p \psi_{i'}^{n+1} \psi_{j'}^{n+1} d\Omega \\
1\le i\le n,\ 1\le j\le n,\ &1\le i'\le n_e,\ 1\le j'\le n_e
\end{aligned} \tag{18}$$

The other coefficient matrices are given similarly.

### 2.3.2 Enforcement of interface temperature condition

The interface temperature condition, i.e. Eq. (5), acts as a constraint condition to the finite element equation, i.e. Eq. (16). This constraint condition was enforced by the penalty method (Chessa et al., 2002). In the penalty method, a constrained problem is simply converted to an unconstrained one. By using Eqs. (10) and (11), Eq. (5) is rewritten as

$$\sum_{i=1}^{n} N_i T_i - \sum_{j=1}^{n_e} N_j \left|\phi_j\right| a_j - T_m = 0 \tag{19}$$

The penalty forcing term is

$$\mathbf{F}_p = \beta \mathbf{G}^T \mathbf{G} \overline{\mathbf{T}}^{n+1} - \beta \mathbf{G}^T \mathbf{c} \tag{20}$$

where $\beta$ is the penalty number,

$$\mathbf{G} = \left[ N_1 \;\; N_2 \;\; \cdots \;\; N_n \;\; -N_1 \left|\phi_1\right| \;\; -N_2 \left|\phi_2\right| \;\; \cdots \;\; -N_{n_e} \left|\phi_{n_e}\right| \right] \tag{21}$$

$$\mathbf{c} = [T_m] \tag{22}$$

and superscript $T$ stands for transpose of the matrix. Adding the global form of the penalty forcing term to the finite element equation, we obtain the final form:

$$\left[ \frac{1}{\Delta t} \mathbf{M}^{n+1} + \mathbf{C}^{n+1} + \mathbf{K}^{n+1} + \beta \left( \mathbf{G}^{n+1} \right)^T \mathbf{G}^{n+1} \right] \overline{\mathbf{T}}^{n+1} = \frac{1}{\Delta t} \mathbf{M}^n \overline{\mathbf{T}}^n + \beta \left( \mathbf{G}^{n+1} \right)^T \mathbf{c} \tag{23}$$

In each of the iterative calculation step to solve Eq. (23), the penalty number is updated to be a value larger than that in the previous step. When the penalty number becomes a large value, $\mathbf{G}\overline{\mathbf{T}}^{n+1} - \mathbf{c}$ is enforced to be small.

It should be noted that the size of the global coefficient matrix changes with move of the solid-liquid interface in the case of multi-dimensional problem and the global coefficient matrix in Eq. (23) is non-symmetric. The Incomplete LU decomposition preconditioned BiConjugate Gradient (ILU-BCG) method was employed to solve Eq. (23).

## 2.4 Gaussian quadrature for enriched elements

The coefficient matrices in the finite element equation, i.e. Eq. (23), are calculated by using the Gaussian quadrature. Since the enrichment function is discontinuous (see Fig. 4), the standard quadrature method will lead to incorrect integration. In order to achieve correct integration in the fully enriched element, the element is divided into two sub-elements as shown in Fig. 5 (Chessa et al., 2002). In the present work, the method to implement the Gaussian quadrature to the sub-elements was formulated by introducing two reference coordinates. Since most problems in nuclear fuel area are multi-dimensional, a numerical technique needs to be applicable to multi-dimensional elements. Although formulation in the case of a two-node line element is described in this section, the method mentioned here can be simply applied to two- or three-dimensional analyses.

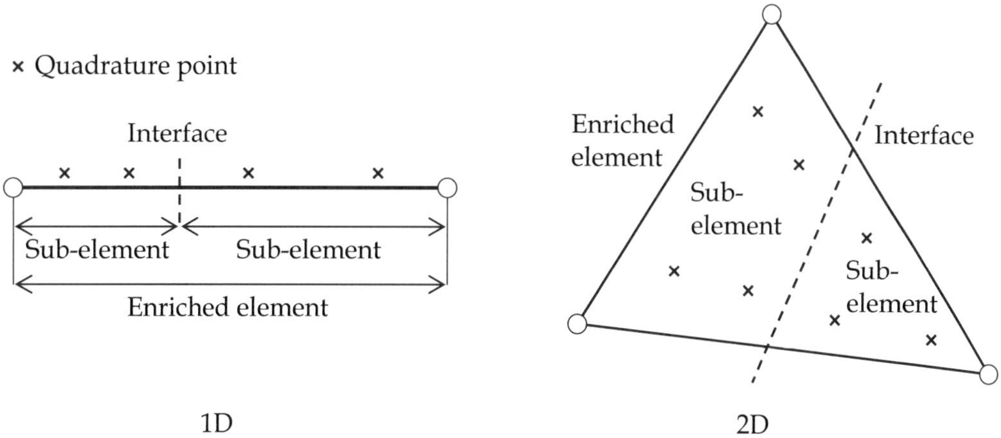


Fig. 5. Sub-elements and quadrature points

The component of the mass matrix **M**, i.e. $\int \rho c_p N_i(x)\psi_j(x)dx$, is divided into two integrations:

$$\int_{\Omega_e} \rho c_p N_i(x)\psi_j^{n+1}(x)dx = \int_{\Omega_{es}} \rho_s c_{ps} N_i(x)\psi_j^{n+1}(x)dx + \int_{\Omega_{el}} \rho_l c_{pl} N_i(x)\psi_j^{n+1}(x)dx \tag{24}$$

where $\Omega_{es}$ and $\Omega_{el}$ stands for the region of the solid- and the liquid-side sub-elements. Here, we introduce a reference coordinate $\xi$ for the enriched element and $\alpha$ for the sub-element (see Fig. 6). Both of the reference coordinate $\xi$ and $\alpha$ range from –1 to 1. By using the Gaussian quadrature, the integration in the region $\Omega_{es}$ is given by

$$\begin{aligned}\int_{\Omega_{es}} \rho_s c_{ps} N_i(x)\psi_j^{n+1}(x)dx &= \rho_s c_{ps}\int_{-1}^{1} N_i(x(\alpha))\psi_j^{n+1}(x(\alpha))J(\alpha)d\alpha \\ &= \rho_s c_{ps}\int_{-1}^{1} N_i(\xi(\alpha))\psi_j^{n+1}(\xi(\alpha))J(\alpha)d\alpha \\ &= \rho_s c_{ps}\sum_{k=1}^{P} N_i(\xi(\alpha_k))\psi_j^{n+1}(\xi(\alpha_k))J(\alpha_k)W_k\end{aligned} \tag{25}$$

where $P$ is the number of the quadrature points, $\alpha_k$ is the coordinate of the $k$-th quadrature point and $W_k$ is the weight at the $k$-th quadrature point. $x(\alpha)$ and $\xi(\alpha)$ in Eq. (25) mean that both of the coordinate $x$ and $\xi$ are the function of the coordinate $\alpha$. $J(\alpha)$ is the Jacobian:

$$J(\alpha) = \frac{\partial x}{\partial \alpha} \tag{26}$$

The integration in the region $\Omega_{el}$ can be obtained similarly. Since calculation of $N_i$ from the coordinate $x$ is almost impossible in multi-dimensional analyses which uses an unstructured mesh, $N_i(x)$ is replaced by $N_i(\xi)$ in Eq. (25). In this case, mapping from $\alpha$ to $\xi$ and mapping from $\alpha$ to $x$ must be given to get the integration. –1 and 1 on the coordinate $\alpha$ is translated into $\xi_1$ and $\xi_I$, respectively. Similarly, –1 and 1 on the coordinate $\alpha$ is translated into $x_1$ and $x_I$, respectively. From this information, the mappings are given by

$$\xi(\alpha) = N_1^{sub}(\alpha)\xi_1 + N_I^{sub}(\alpha)\xi_I \tag{27}$$

$$x(\alpha) = N_1^{sub}(\alpha)x_1 + N_I^{sub}(\alpha)x_I \tag{28}$$

where $N_1^{sub}(\alpha)$ and $N_I^{sub}(\alpha)$ are the shape functions shown in Fig. 6. The right hand side of Eqs. (27) and (28) are computable. We can calculate $\xi(\alpha_k)$ and $J(\alpha_k)$ from Eqs (26) to (28), and obtain the correct integration.

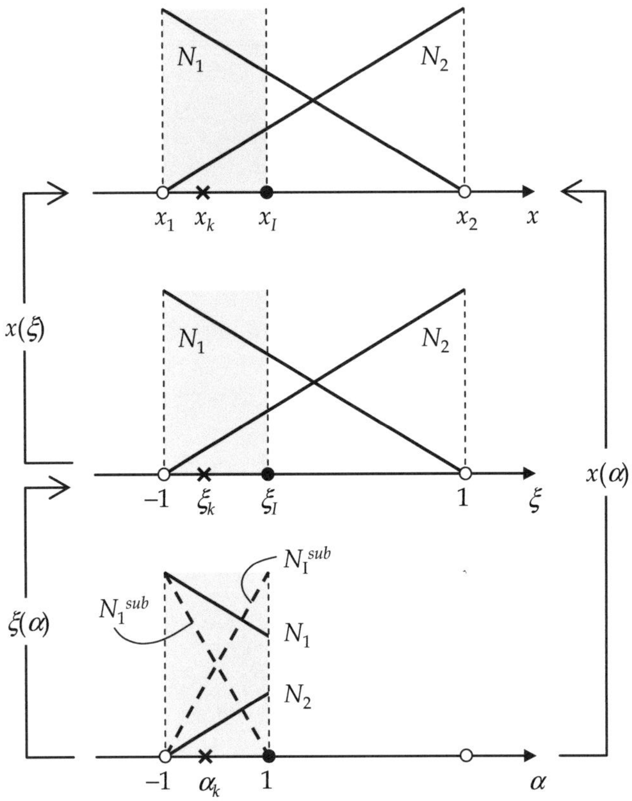


Fig. 6. Reference coordinates for enriched element

## 2.5 Update of signed distance function

The signed distance function is updated by the algorithm of the level set method (Osher & Sethian, 1988; Sethian, 1999; Chessa et al., 2002). In this method, interface motion is described by the advective equation:

$$\frac{\partial \phi}{\partial t} + F|\nabla \phi| = 0 \tag{29}$$

where $F$ is the extension velocity to be described hereinafter. The Galerkin/Least Squares (GLS) method (Hughes et al., 1989) was applied to Eq. (29) for stabilization. The stabilized finite element equation is

$$\int_{\Omega}\delta\phi\frac{\partial\phi}{\partial t}d\Omega+\int_{\Omega}\delta\phi F|\nabla\phi|d\Omega+\sum_{e}\int_{\Omega_e}\left(\frac{F}{|\nabla\phi|}\nabla\delta\phi\cdot\nabla\phi\right)\tau_e\left(\frac{\partial\phi}{\partial t}+F|\nabla\phi|\right)d\Omega=0 \tag{30}$$

where $\delta\phi$ is the weight function and $\tau_e$ is the stabilization parameter. Formulation of the stabilization parameter was proposed by Tezduyar (1992). To solve Eq. (30), we first have to construct the extension velocity, which advects the signed distance function. The extension velocity has the characteristic that it be orthogonal to the signed distance function:

$$\mathrm{sign}(\phi)\nabla F\cdot\nabla\phi=0 \tag{31}$$

Figure 7 shows the essential boundary condition for the extension velocity. S is the intersection point between the solid-liquid interface and a perpendicular line from the enriched node P. Considering the orthogonal condition given by Eq. (31), it is obvious that the extension velocity at the node P is equal to the interface velocity at S. Equation (31) is solved with taking the extension velocity at the enriched node as an essential boundary condition. The stabilized form of Eq. (31) is written as

$$\int_{\Omega}\delta F\mathrm{sign}(\phi)\nabla F\cdot\nabla\phi d\Omega+\sum_{e}\int_{\Omega_e}(\nabla\delta F\cdot\nabla\phi)\tau_e(\nabla\phi\cdot\nabla F)d\Omega=0 \tag{32}$$

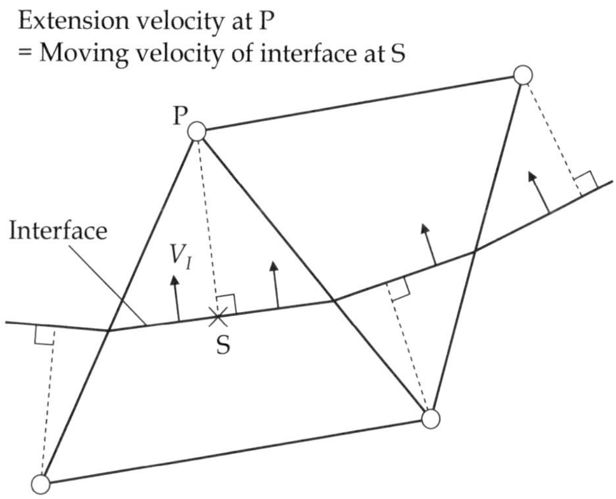


Fig. 7. Essential boundary condition of extension velocity

### 2.6 Calculation of liquid flows

Most problems in nuclear fuel area involve liquid flow. Liquid flow plays an important role in determining the shape of the solid-liquid interface. For example in injection casting or the process of high-level radioactive wastes disposal, flow of the molten mixture has an effect on performance of their process. In actual application of the numerical analysis, liquid flow and its effect on the behavior of the interface must be considered.

In our method, Eqs. (1) and (2) are solved by the velocity correction method (Ramaswamy, 1988). The Streamline-Upwind/Petrov-Galerkin (SUPG) scheme (Brooks & Hughes, 1982) was applied to the convective term in Eq. (2). The effect of a buoyancy force was modelled by the Boussinesq approximation. To consider the condition that the flow velocity is 0 at the solid-liquid interface, the following essential boundary condition was applied to the enriched nodes which exist in the solid region:

$$\mathbf{u}_{ebc} = -\frac{\phi_l}{\phi_s}\bar{\mathbf{u}} \tag{33}$$

where $\mathbf{u}_{ebc}$ is the velocity at the enriched node in the solid region, $\phi$ is the length of the perpendicular line from the enriched node to the solid-liquid interface and $\bar{\mathbf{u}}$ is the velocity at the intersection between the perpendicular line and the element boundary (see Fig. 8). Equation (33) limits the velocity to 0 at the solid-liquid interface. $\mathbf{u}_{ebc}$ approaches 0 as the solid-liquid interface comes close to the enriched node in the solid region.

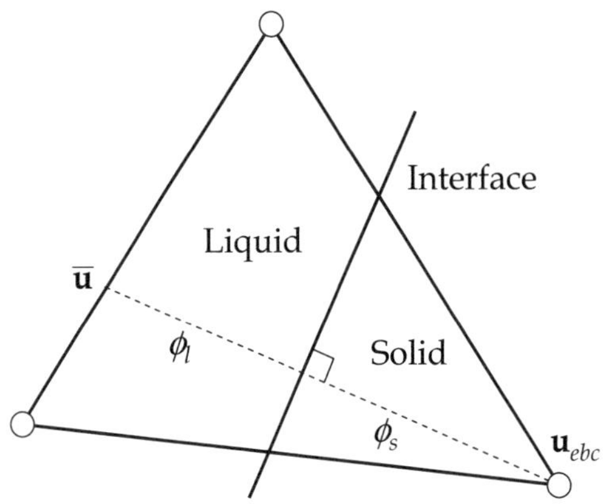


Fig. 8. Essential boundary condition of liquid-phase velocity

### 2.7 Time-stepping procedure

The time-stepping procedure is summarized below.

1. Construct the extension velocity $F$ from Eq. (32) under the essential boundary condition shown in Fig. 7. Equation (4) is used to give the essential boundary condition.
2. Calculate the signed distance function $\phi^{n+1}$ from Eq. (30).
3. Calculate the flow velocity $\mathbf{u}^{n+1}$ from Eqs. (1) and (2) under the essential boundary condition given by Eq. (33).
4. Build the global matrix in Eq. (23) and solve for $T^{n+1}$ by using the penalty method.
5. Return to step 1 and repeat the algorithm.

## 3. Results and discussion

### 3.1 One-dimensional Stefan problem

Numerical analysis of a one-dimensional melting problem using the X-FEM is now presented. The temperature at the one boundary is set to a value above the melting temperature, and the temperature at the other boundary is set to the melting temperature. This boundary condition keeps the temperature to the melting temperature in the whole solid region and causes move of the solid-liquid interface. This problem is the well-known one-phase Stefan problem. The exact solution describing the interface motion is given by

$$x_I(t) = 2\eta\sqrt{\alpha_l t} \tag{34}$$

where $x_I(t)$ is the interface position and $\alpha_l$ is the thermal diffusivity of the liquid phase. The constant $\eta$ satisfies the relationship:

$$\frac{e^{-\eta^2}}{\operatorname{erf}(\eta)} = \frac{\eta L\sqrt{\pi}}{c_l(T_{lw} - T_m)} \tag{35}$$

The temperature profile in the liquid region is given by

$$T_l(x,t) = T_{lw} - \frac{T_{lw} - T_m}{\operatorname{erf}(\eta)}\operatorname{erf}\left(\frac{x}{2\sqrt{\alpha_l t}}\right) \tag{36}$$

The region which is 0.02 m in length was divided into the 10 finite elements. The initial temperature profile was given by the exact solution at $t$ = 200 sec (i.e. the solid-liquid interface was initially at $x = 2.54\times10^{-3}$ m). The boundary conditions were $T_{lw}$ = 10 °C at $x$ = 0 m and $T_{sw}$ = 0 °C at $x$ = 0.02 m. The physical properties of the water were used in this analysis.

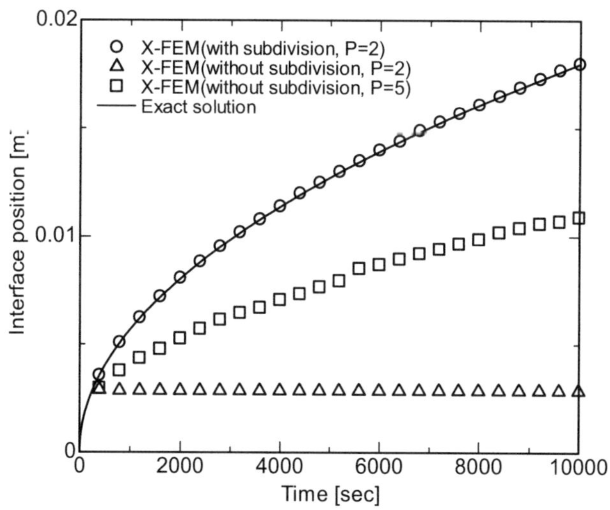


Fig. 9. Comparison of calculated interface position with exact solution in one-dimensional Stefan problem

Figure 9 shows the calculated interface positions and the exact solution as a function of time. The numerical analysis without subdivision of the enriched element was also conducted and its results are shown for comparison. The number of the quadrature point $P$ was taken as a parameter. As can be seen, the numerical analysis using the subdivision technique predicted the interface motion successfully. The relative error is within 1 %. When subdivision was not applied, the numerical analysis could not reproduce the interface motion even in the case of $P = 5$. Figure 10 shows the temperature profiles at the four different times. The calculated temperature profiles agree with the exact solutions excellently. From this verification, it has been demonstrated that the X-FEM could predict the interface motion and the temperature profile accurately even in a fixed mesh.

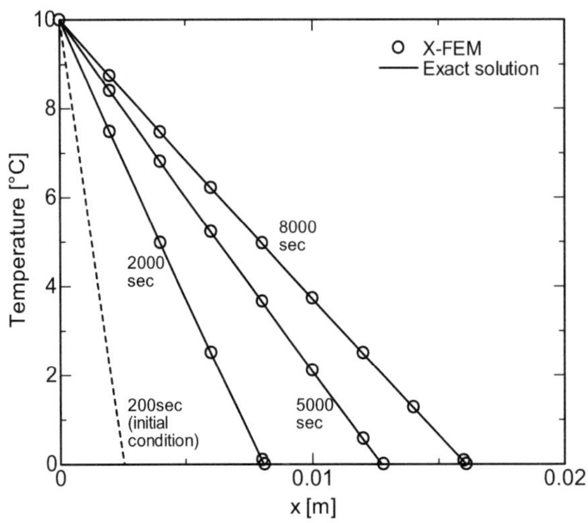


Fig. 10. Comparison of calculated temperature profile with exact solution in one-dimensional Stefan problem

### 3.2 Solidification in two-dimensional square corner

The problem of solidification in a square corner was analyzed as a part of verification to multi-dimensional problems. An infinitely long prism is initially filled with a fluid at its melting temperature. Figure 11 shows the square-shaped cross section of the prism. The temperature on the peripheral surface of the prism is maintained at a certain value below the melting temperature so that solidification proceeds from the surface inward. There are no changes of the physical properties, and hence there is no convection in the liquid region. As shown in Fig. 11, the verification analysis was performed for the quarter region by taking the centerline of the prism as an adiabatic boundary. The analysis region was divided into the 400 triangle elements. The physical properties of the water were used in this analysis.

Figure 12 shows the temperature profile and the shape of the solid-liquid interface at the different times. The solid-liquid interface is denoted by the isoline of $\phi = 0$. During the initial periods of solidification, the shape of the interface is close to square. As the time progresses, the curve near the diagonal gradually flattens. The interface position along the centerline, i.e. the adiabatic boundary, and along the diagonal line is shown against time in Fig. 13, respectively. The approximate solutions derived by Allen & Severn (1962), Crank & Gupta (1975), Lazaridis (1970) and Rathjen & Jiji (1971) are plotted for comparison. The results of the X-FEM agree with some of the approximate solutions very well. The difference between the numerical results and some of the solutions seems to be due to the approximations used in the previous researches.

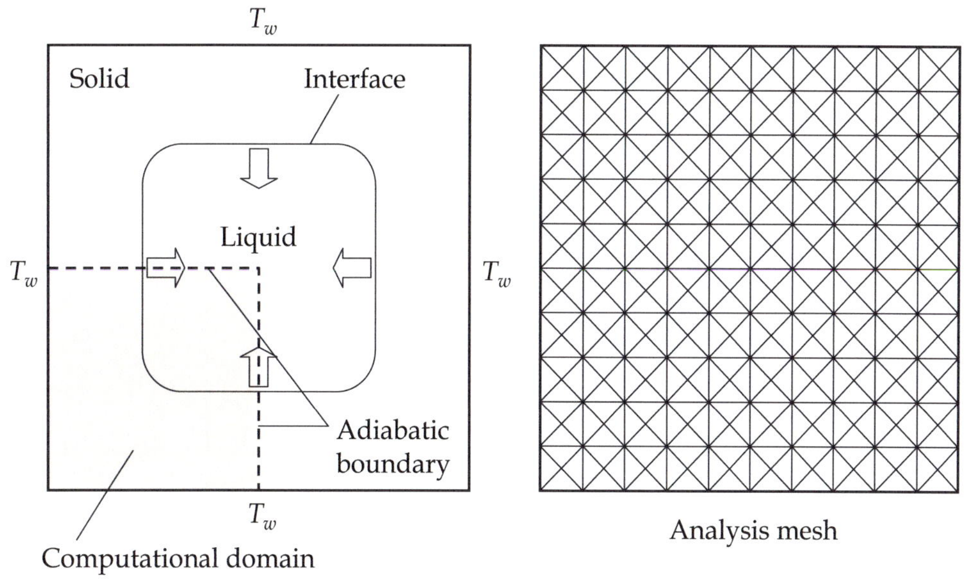


Fig. 11. Computational domain for solidification in two-dimensional square corner

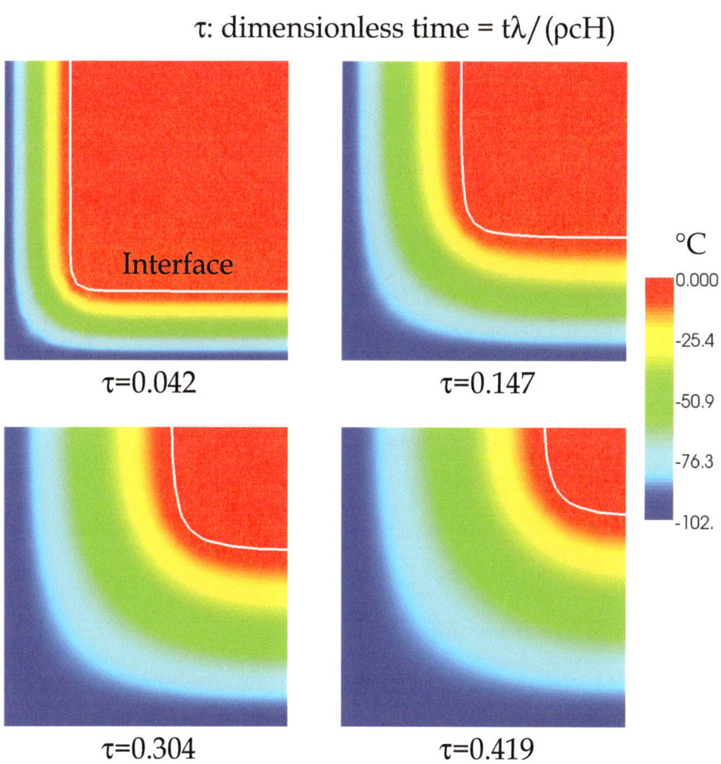


Fig. 12. Calculated solid-liquid interface and temperature in two-dimensional square corner

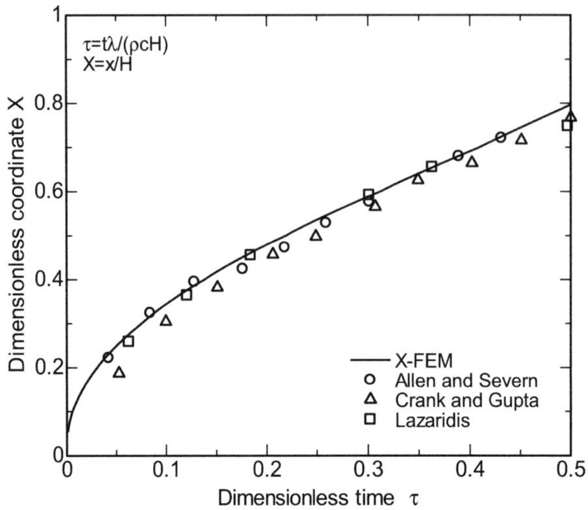


(a) Interface position along center line

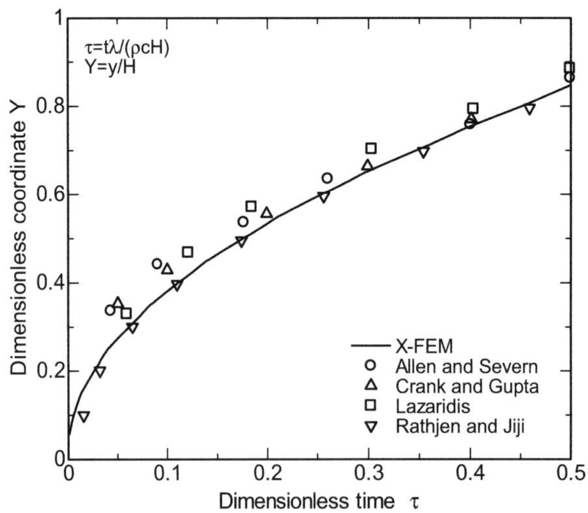


(b) Interface position along diagonal line

Fig. 13. Comparison of calculated interface position with existing solutions in two-dimensional square corner

### 3.3 Melting of gallium in two-dimensional cavity

Gau & Viskanta (1986) conducted the experiment on melting of pure gallium. Their experiment was analyzed to confirm applicability of our numerical analysis method to the problem involving liquid flow. The two-dimensional cavity which is 0.04445 m in height and 0.0889 m in width (see Fig. 14) is initially filled with pure solid gallium at its melting temperature (29.78 °C). At $t = 0$, the temperature on the left-hand wall is raised to 38 °C. On the other hand, the temperature on the right-hand wall is kept at the melting temperature. The upper and the lower walls are insulated. As time increases, melting proceeds from the left-hand vertical wall rightward. In this problem, natural convection occurs because of the existence of the gravitational force and the density change. The cavity was divided into the 800 triangle elements.

Figure 15 shows the calculated interface, temperature profile and streamlines. It can be seen that the solid-liquid interface inclined rightward from the vertical axis. Since the temperature gradient near the bottom of the liquid region is relatively low, the interface at the bottom moves more slowly. Thus, natural convection plays an important role in determining the shape of the solid-liquid interface. As shown in Fig. 16, the calculated interface shape shows good agreement with the experimental results by Gau & Viskanta. Applicability to the actual problems in the nuclear fuel area was demonstrated through this analysis. The slight difference is probably due to the uncertainty on the temperature at the right-hand wall in the experiment, which is mentioned by Gau & Viskanta (1986), and the anisotropic thermal conductivity of the pure gallium (Cubberley, 1979).

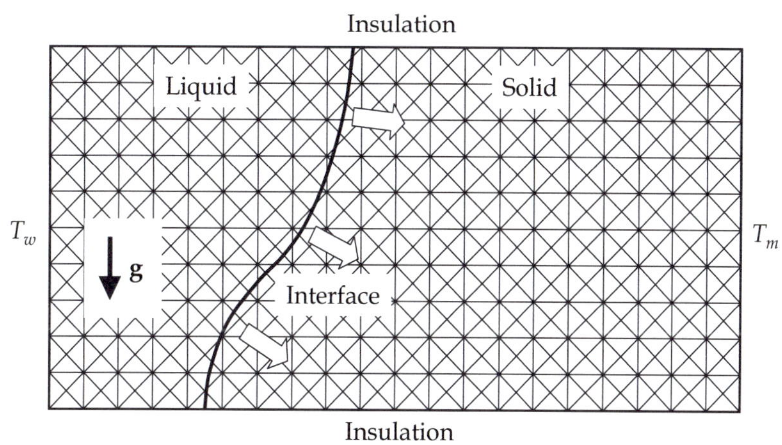


Fig. 14. Computational domain for melting of pure gallium in two-dimensional cavity

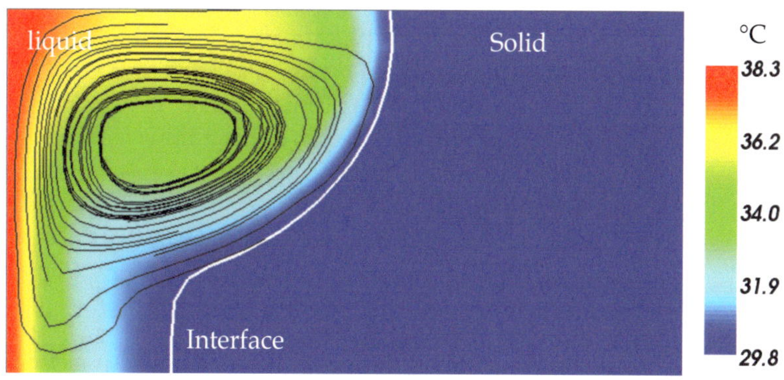


Fig. 15. Calculated solid-liquid interface, temperature and streamlines in two-dimensional gallium melting problem

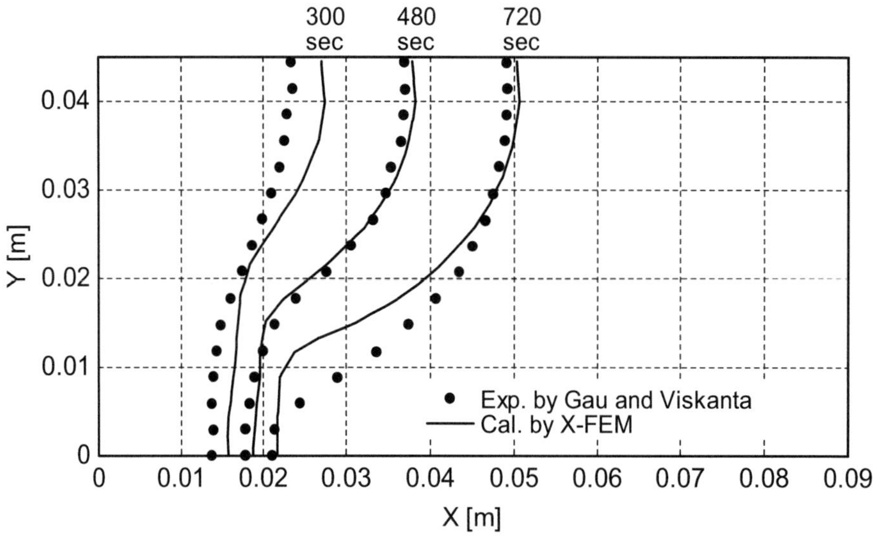


Fig. 16. Comparison of calculated interface position with experimental data in two-dimensional gallium melting problem

## 4. Conclusion

In this study, the X-FEM was applied to melting/solidification problems. The method uses the enriched finite element interpolation to represent the discontinuous temperature gradient in the element crossed by the solid-liquid interface. Construction of the enriched finite element interpolation and the finite element equation using it was discussed. We formulated the numerical integration for the enriched element. The method to solve problems involving liquid flows was also developed in the present work. From the numerical analysis of the one-dimensional Stefan problem, it has been demonstrated that the X-FEM could predict the interface motion and the temperature profile accurately even in a fixed mesh. As a part of verification to multi-dimensional problems, the solidification in the two-dimensional square corner was also analyzed. The numerical results on the interface position showed good agreement with some of the approximate solutions. In the analysis of melting of pure gallium, the shape of the solid-liquid interface was reproduced successfully. These verifications indicate that our numerical analysis method is physically correct and applicable to the multi-dimensional problems involving liquid flow. This method makes it possible to evaluate melting/solidification processes appearing in injection casting and the process of high-level radioactive wastes disposal, etc. The future work is to develop a mass transfer model to simulate multi-component dissolution/precipitation phenomena.

## 5. References

Allen, D. N. de G. & Severn, R. T. (1962). The Application of Relaxation Methods to the Solution of Non-elliptic Partial Differential Equations III. *Q. Jl Mech. Appl. Math.*, Vol.15, pp. 53-62, 1962.

Brooks, A. N. & Hughes, T. J. R. (1982). Streamline-Upwind/Petrov-Galerkin Formulations for Convection Dominated Flows with Particular Emphasis on the Incompressible Navier-Stokes Equations. *Comput. Meth. Appl. Mech. Engrg.*, Vol.32, pp. 199, 1982.

Chessa, J. & Belytschko, T. (2003). The Extended Finite Element Method for Two-phase Fluids. *ASME J. Appl. Mech.*, Vol.70, pp. 10-17, 2003.

Chessa, J.; Smolinski, P. & Belytschko, T. (2002). The Extended Finite Element Method (X-FEM) for Solidification Problems. *Int. J. Numer. Meth. Engng.*, Vol.53, pp. 1959-1977, 2002.

Chessa, J.; Wang, H. & Belytschko, T. (2003). On Construction of Blending Elements for Local Partition of Unity Enriched Finite Element Methods. *Int. J. Numer. Meth. Engng.*, Vol.57, pp. 1015-1038, 2003.

Comini, G.; Guidice, S. Del; Lewis, R. W. & Zienkiewicz, O. C. (1974). Finite Element Solution on Non-Linear Heat Conduction Problems with Special Reference to Phase Change. *Int. J. Numer. Meth. Engng.*, Vol.8, pp. 613-624, 1974.

Crank, J. & Gupta, R. S. (1975). Isotherm Migration Method in Two Dimensions. *Int. J. Heat Mass Transfer*, Vol.18, pp. 1101-1107, 1975.

Cubberley, W. H. Metal Handbook–Properties and Selection: Nonferrous Alloys and Pure Metals, 9th edn., American Society of Metals, Metals Park, OH, pp. 736-737, 1979.

Funasaka, H. & Itoh, M. (2007). Perspective and Current Status on Fuel Cycle System of Fast Reactor Cycle Technology Development (FaCT) Project in Japan. *Proceedings of the GLOBAL 2007*, Boise, Idaho, USA, Sep 9–13, 2007.

Gau, C. & Viskanta, R. Melting and Solidification of a Pure Metal on a Vertical Wall. *Trans. ASME J. Heat Transfer*, Vol.108, pp. 174-181, 1986.

Hughes, T. J. R.; Franca, L. P. & Hulbert, G. M. (1989). A New Finite Element Formulation for Computational Fluid Dynamics: VIII. The Galerkin/Least-Squares Method for Advective-Diffusive Equations. *Comput. Meths. Appl. Mech. Engrg.*, Vol.73, pp. 173-189, 1989.

Jaeger, M. & Carin, M. (1994). The Front-Tracking ALE Method: Application to a Model of the Freezing of Cell Suspensions. *J. Comp. Phys.*, Vol.179, pp. 704-735, 1994.

Kofuji, H. (2010). Electro-Deposition Behavior of Minor Actinides with Liquid Cadmium Cathodes. *IOP Conference Series; Materials Science and Engineering*, vol.9, pp. 012010-1-012010-8, 2010.

Kozaka, T. & Tominaga, S. (2005). Development of the Vitrification Technology at TVF. *JNC Technical Review*, JNC TN1340 2005-002, pp. 35-39, 2005.

Lazaridis, A. (1970). A Numerical Solution of the Multidimensional Solidification (or Melting) Problem. *Int. J. Heat Mass Transfer*, Vol.13, pp. 1459-1477, 1970.

Lynch, D. R. & O'Neill, K. (1981). Continuously Deforming Finite Elements for the Solution of Parabolic Problems, with and without Phase Change. *Int. J. Numer. Meth. Engng.*, Vol.17, pp. 81-96, 1981.

Merle, R. and Dolbow, J. (2002). Solving Thermal and Phase Change Problems with the Extended Finite Element Method. *Comp. Mech.*, Vol.28, pp. 339-350, 2002.

Moës, N.; Dolbow, J. & Belytschko, T. (1999). A Finite Element Method for Crack Growth without remeshing. *Int. J. Numer. Meth. Engng.*, Vol.46, pp. 131-150, 1999.

Murray, W. D. & Landis, F. (1959). Numerical and Machine Solutions of Transient Heat-Conduction Problems Involving Melting or Freezing. *ASME J. Heat Transfer*, Vol.81, pp. 106-112, 1959.

Osher, S. & Sethian, J. A. (1988). Propagation of Fronts with Curvature Based Speed: Algorithms Based on Hamilton-Jacobi Formulations. *J. Comp. Phys.*, Vol.79, pp. 12-49, 1988.

Ramaswamy, B. (1988). Finite Element Solution for Advection and Natural Convection Flows, *Comput. Fluids*, Vol.16, pp. 349, 1988.

Rathjen, K. A. & Jiji, L. M. (1971). Heat Conduction with Melting or Freezing in a Corner. *Trans. ASME J. Heat Transfer*, Vol.93, pp. 101-109, 1971.

Rolph III, W. D. & Bathe, K. J. (1982). An Efficient Algorithm for Analysis of Nonlinear Heat Transfer with Phase Change. *Int. J. Numer. Meth. Engng.*, Vol.18, pp. 119-134, 1982.

Sagayama, Y. (2007). Launch of Fast Reactor Cycle Technology Development Project in Japan. *Proceedings of the GLOBAL 2007*, Boise, Idaho, USA, Sep 9–13, 2007.

Sampath, R. & Zabaras, N. (1999). An Object Oriented Implementation of a Front Tracking Finite Element Method for Directional Solidification Processes. *Int. J. Numer. Meth. Engng.*, Vol.44, pp. 1227-1265, 1999.

Sethian, J. A. (1999). Level Set Methods and Fast Marching Methods. Evolving Interfaces in Computational Geometry, Fluid Mechanics, Computer Vision, and Materials Sciences. *Cambridge Monographs on Applied and Computational Mathematics*, Cambridge University Press, Cambridge, 1999.

Tezduyar, T. E. (1992). Stabilized Finite Element Formulations for Incompressible Flow Computations. *Adv. Appl. Mech.*, Vol.28, pp. 1-44, 1992.

# 4

# Relativistic Density – Functional Study of Nuclear Fuels

Masayoshi Kurihara and Jun Onoe
*Research Laboratory for Nuclear Reactors and Department of Nuclear Engineering, Tokyo Institute of Technology, Tokyo Japan*

## 1. Introduction

Among nuclear materials used in both research and commercial power reactors, very high-density dispersion fuels have been used for requirement of a large proportion of uranium per unit volume in order to compensate to the reduction of enrichment (Meyer et al., 2002, Kim et al., 1999, 2002). For examples, uranium (U) with 10 wt% molybdenum (Mo) dispersed in aluminum (Al) matrix (Meyer et al., 2002, Kim et al., 2002) and $U_3Si_2$/Al dispersion fuel (Kim et al., 1999) have been hitherto examined. Furthermore, zirconium (Zr) based U/plutonium (Pu) alloys have recently been focused as a promising fuel for advanced reactors (Chernock & Horton, 1994).

U metals have α (orthorhombic)-, β (tetragonal)-, and γ (body centered cubic)-phases that depend on the ambient temperature (Chiotti et al., 1981). The α and β U phases form solid solutions with other metal elements to a limited extent (Chiotti et al., 1981), whereas the γ-U phase forms solid solutions with other elements to any extent (Chiotti et al., 1981). In addition, the α-U phase forms many intermetallic compounds with other metal elements (Kaufman, 1961). In these reasons, some α- and/or γ-U/transition metal (TM) alloys have been used in research reactors (Chiotti et al., 1981, Kaufman, 1961). To use advanced nuclear fuels practically, it is critical to understand the dissolution process of TM atoms into γ-U on the basis of their phase diagrams.

The alloying behavior of U compounds has been hitherto theoretically investigated by considering the relative stabilities of electronic configurations (Buzzard, 1955, Park & Buzzard, 1957) and the thermodynamics (Ogawa et al., 1995). Hume-Rothery has focused on the metallic radius of TM elements as a parameter for understanding their alloying behavior (Hume-Rothery & Raynor, 1954). According to their empirical findings, when the difference in the metallic radius between solute and solvent atoms is less than 15%, TM elements are very soluble to each other. Figure 1 shows the correlation between U/TM radius and the maximum solid solubility (MSS) of TM into γ-U. Here, the dashed line denotes the metal radius smaller by 15% than that of γ-U (Zachariasen, 1973, Pauling, 1960).

This rule seems to explain MSS for 3d-TM/γ-U alloys, but cannot explain MSS for 4d- and 5d-TM/γ-U alloys. Furthermore, Buzzard pointed out that the miscibility of TMs into γ-U depends on compatibility factors considering ionic radius, electron negativity, and the number of d-electrons (Buzzard, 1955, Park & Buzzard, 1957). However, because many empirical parameters should be determined, this evaluation process of their alloying behaviors is not applicable to other actinide alloy systems. On the other hand, Ogawa et al. investigated the alloying behavior by using the ChemSage program, and concluded that the difference in the excess free energy ($\Delta G^E$) of the U/Mo, U/rhodium (Rh), and U/palladium (Pd) alloying systems may be due to the contribution of the U6d-Pd4d orbital interactions (Ogawa et al., 1995). Although the interactions indeed play an important role of their alloying behaviors, it is necessary to clarify what interactions between γ-U and TMs contribute to the alloying behavior quantitatively in order to satisfactorily understand the alloying behavior of the 3d, 4d and 5d TMs into γ-U solid.

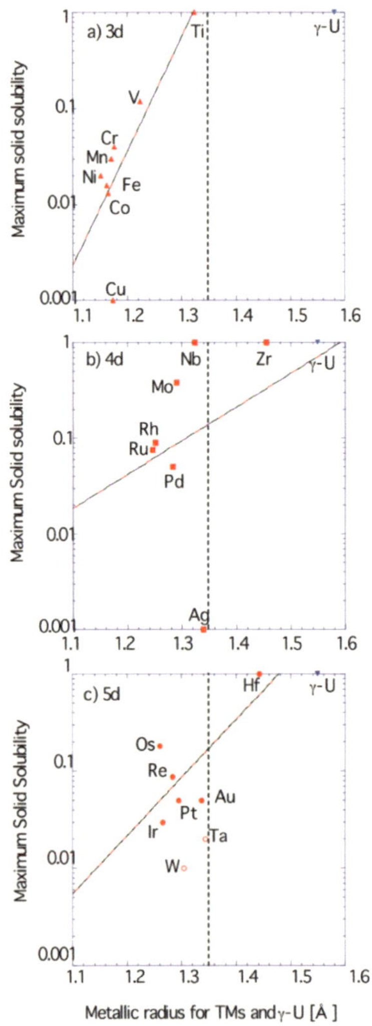


Fig. 1. The correlation between the metallic radius of TMs and the maximum solid solubility for γ-U alloyed with (a) 3d TMs, (b) 4d TMs, and (c) 5d TMs [Kurihara et al., (2008)]

Recently, some groups have theoretically studied the thermodynamic properties of U/TM alloys. Landa et al. have examined the phase equilibrium of U/Zr alloys, using the scalar-relativistic Green's function technique based on Korrings-Kohn-Robstoker (KKR) method (i.e. without spin-orbit coupling) (Landa et al., 2009). They obtained a good agreement between theoretical and experimental results for the ground-state properties of γ- (bcc) and δ- (C32) phases of U/Zr alloys. On the other hand, Li et al. have obtained the thermodynamic assessments for thorium (Th)/U and Th/Zr binary and Th/U/Zr ternary alloys (Li et al., 2009), using the CALPHAD (calculation of phase diagrams) method based on experimental data including the phase equilibria and thermodynamic properties of the alloys. They have also obtained a good agreement between the calculated phase equilibria and experimental data. *Ab initio* calculations based on density-functional theory have also been used to examine the thermodynamic properties of U/Al systems (Alonso et al., 2009, Sedmidudbsky et al., 2010). To understand the dissolution of elements into solid U in addition to obtain the phase diagram or phase equilibria, it is indespensable to determine which electronic terms contribute to the thermodynamic properties of U alloys by comparing density-functional calculations with experimental results.

We have previously investigated the alloying behavior of 3d-, 4d- and 5d-TMs into γ-U, using the discrete-variational Dirac-Fock-Slater molecular orbital (MO) method (R-DFT) which takes into account fully relativistic effects including the spin-orbit coupling (Kurihara et al., 2004, 2008, 2011). We have found that the maximum solid solubility (MSS) of TMs into γ-U is exponentially proportional to both the d-orbital energy of TM (Md) and the orbital overlap population (OOP) between TMd and U6d atomic orbitals.

In this chapter, we firstly introduce the relativistic discrete-variational Xα molecular orbital method (R-DFT) (Rosen et al., 1975, Adachi et al., 1977, Nakamatsu et al., 1991), secondary introduce our recent works on analysis of the valence photoelectron spectra of uranium carbides (UC) and α-U metal, and thirdly show the elucidation of the alloying behavior of TMs into γ-U metal using the R-DFT, and finally summarize the results obtained using R-DFT and describe the perspective for the quantum design of nuclear fuels for advanced reactors.

## 2. Computational method

### 2.1 Relativistic density-functional method

The present R-DFT method has been shown to be a powerful tool for the study of the electronic structures of molecules containing heavy elements such as uranium (Onoe et al., 1992a, 1994b, Hirata et al., 1997a, Kurihara et al., 1999, 2000). The one-electron molecular Hamiltonian, *H*, in the R-DFT method is written as,

$$\boldsymbol{H} = c\boldsymbol{\alpha}\boldsymbol{P} + \boldsymbol{\beta}mc^2 + V(r)\,. \quad (1)$$

Here, $c$, $\boldsymbol{P}$, $m$, $\boldsymbol{\alpha}$, $\boldsymbol{\beta}$ and $V(r)$ respectively denote the velocity of light, the operator of momentum, the mass of electron, Dirac matrices and the sum of Coulomb and exchange potentials. The molecular orbitals (MOs) are obtained by taking a linear combination of atomic orbitals (AOs). Details of the R-DFT method have been described elsewhere (Rosen et al., 1975, Onoe et al., 1993).

Basis functions, which are numerical solutions of the atomic Dirac-Fock-Slater equations for an atomic-like potential, are obtained at the initial stage of individual iterations of self-consistent procedures (Adachi et al., 1977). The atomic-like potentials used to generate the basis functions are derived from the spherical average of the molecular charge density around nuclei. One-center (atomic) charges are estimated using Mulliken population analysis (Mulliken et al., 1955a, 1955b, 1955c, 1995d) for application to the self-consistent charge (SCC) method (Rosen et al., 1976) that is used to approximate the self-consistent field.

Morinaga et al have first found that the d-electron energy (Md) plays an important role for the alloying behavior (Morinaga et al., 1984, 1985a, 1985b, 1985c, 1991, 2005). In a similar manner, we have evaluated the Md of TMs in γ-U/TM alloys from a weighted average of each component ($d_{3/2}$ and $d_{5/2}$), along with the U6d orbital energy. To compare Md values obtained for all TMs and U with each other, Md was shifted with respect to the Fermi level ($E_F$) of γ-U used as a reference (Morinaga et al., 1984, 1985a).

Since the orbital overlap population (OOP) expresses the strength of covalent interactions between AOs, it is a powerful tool to clarify the contribution of individual AOs to covalent bonding (Onoe et al., 1997c). In order to obtain information of the covalent bonding, we employed the Mulliken population analysis (Mulliken et al., 1955a, 1955b, 1955c, 1995d). In the present analysis, the number of electrons ($n_i$) was partitioned into the gross of the $i$th AO,

$$n_i = \sum_{(l,j)} \phi_l C_{il} C_{jl} S_{ij}, \tag{2}$$

Here, $\phi_l$ is the occupancy of the $l$th MO, $C_{il}(C_{jl})$ is the coefficient of the linear combination of AOs, and $S_{ij}$ is the overlap integral between the $i$th and $j$th AOs. Two-center charges were estimated from the overlap populations ($i$th AO of γ-U and $j$th AO of TM),

$$n_{ij} = \sum_{(l)} \phi_l C_{ij} C_{jl} S_{ij}. \tag{3}$$

The effective charges for γ-U/TM alloys were evaluated, because they are strongly related to the charge transfer (CT) between TM and γ-U.

### 2.2 Cluster model

Unlike band structure calculations of condensed matters obtained using their unit cells with a periodic boundary condition, the present R-DFT method employed a cluster model reflecting the crystal structure.

#### 2.2.1 Uranium carbide

For a cluster model of uranium carbides (UC) with a NaCl-type structure, we made neutral $UC_6$, $CU_6$, and $CU_6C_{18}$ cluster models with Oh symmetry and with a U-C bond length of 248.1 pm taken from the experimental results (Erode, P., 1983), as shown in Fig. 2.

In case of the $CU_6C_{18}$ cluster, the $CU_6$ cluster was embedded with eighteen carbon atoms. Since the spin function is included in the Dirac equation explicitly, the Oh symmetry reduces to the Oh* double group (Bethe, H., 1929). Symmetry orbitals for the irreducible representations of Oh* symmetry were constructed from AOs by the projection operator method (Meyer, J. et al., 1989). All the calculations were performed with the Slater exchange parameter α of 0.7. The DV sample points of 6000 were used for the $UC_6$ and $CU_6$ models, while 16,000 points for the $CU_6C_{18}$ model. The basis functions up to the 7p orbital were used for U atom and those up to the 2p orbital were used for C atom. All the calculations were carried out self-consistently until the difference between the initial and final orbital populations in the iteration was less than 0.01.

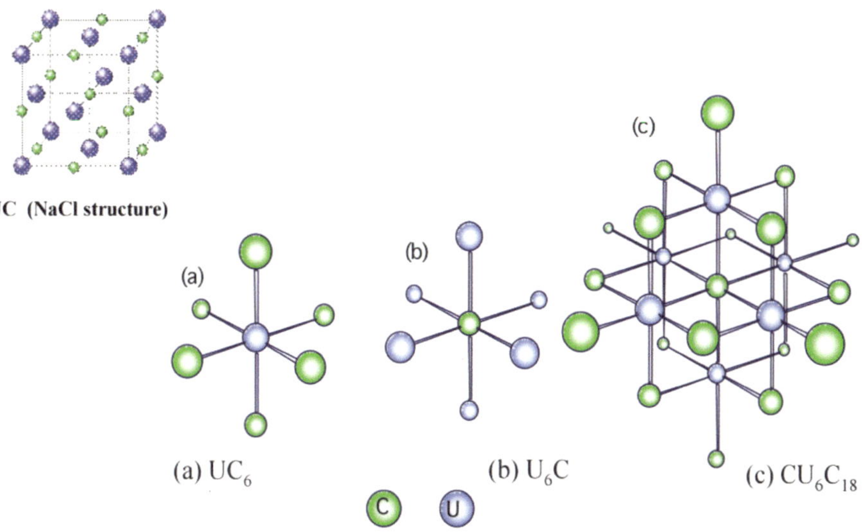


Fig. 2. Schematic illustration of cluster models for UC with a NaCl-type crystal structure: (a) $UC_6$, (b) $CU_6$, and (c) $CU_6C_{18}$ [Kurihara et al., (1999)]

### 2.2.2 α-Uranium metal

Figure 3 shows (a) the unit cell of α-U metal crystal with an orthorhombic structure and (b) the $U_9$ cluster model with the central U atom (1) surrounded by eight U atoms (2)-(8).

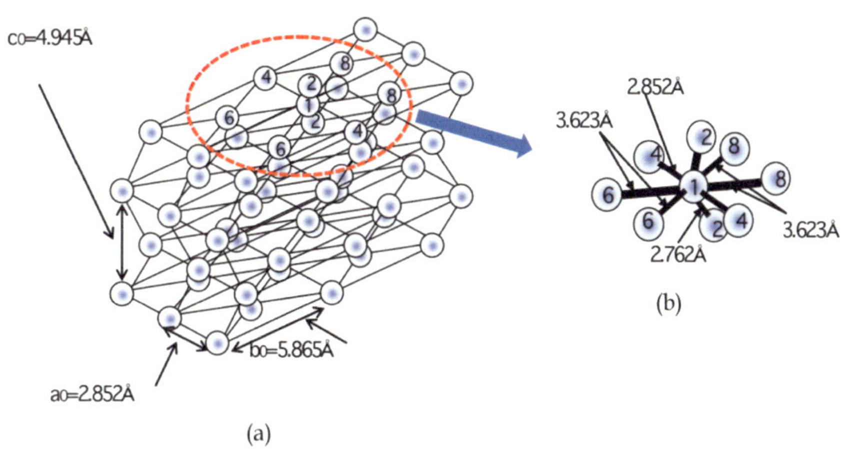


Fig. 3. Schematic illustration of α-U metal: (a) the orthorhombic structure of α-U metal and (b) a $U_9$ cluster model used as the minimum unit of α-U crystal structure [Kurihara et al., (2000)].

The geometry of the present cluster has a $C_{2v}$ symmetry with the U-U bond length of 5.219 au for (1)-(2), 5.389 au for (1)-(4), and 6.165 au for (1)-(6) and (1)-(8) on the basis of the experimental results of α-U metal (Holden, A. N., 1958). Here, "au" denotes the atomic unit (1 au = Bohr radius). Symmetry orbitals for the irreducible representations of $C_{2V}^*$ symmetry were constructed from AOs by the projection operator method. All the calculations were performed with the Slater exchange parameter α of 0.7. The DV sample points of 18,000 were used for the $U_9$ cluster model. The basis functions up to the 7p orbital were used for U atom. All the calculations were carried out under the same conditions as for UC.

### 2.2.3 γ-Uranium/transition metal alloy

Figure 4 shows schematic representation of a cluster model of γ-U alloyed with TMs. As shown in Fig. 4, the central U atom, U(1), was substituted with TM atoms. To compare the difference in the electronic structure among the γ-U/TM alloys, the MO energy level structure and chemical bonding of the alloys were examined (Kurihara et al., 2004). In the present works, the lattice relaxation in association with TM element substitution to γ-U solid was ignored, because the Md and OOP may not be affected significantly by the lattice relaxation (Morinaga et al., 1984, 1985a, 1985b). The lattice constant of γ-U crystal was taken from the experimental result of 6.659 au (Holden 1958).

Symmetry orbitals for the irreducible representations of $D_{4h}^*$ symmetry were constructed from the AOs by the projection operator method. The present R-DFT calculations were performed with the Slater exchange parameter α of 0.7 and with 18,000 DV sample points for the $U_8$-TM cluster model. The basis functions were used to be the 1s-7p AOs for U atom, the 1s-4p AOs for 3d TMs, the 1s-5s AOs for 4d TMs, and the 1s-6s AOs for 5d TMs. All the calculations were carried out self-consistently under the same conditions as for UC and α-U.

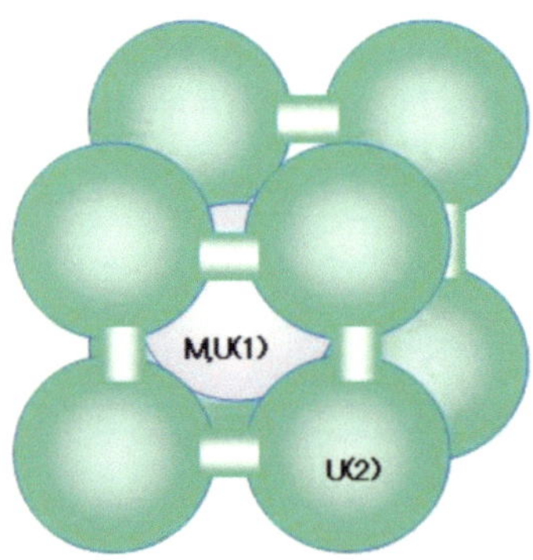


Fig. 4. Schematic illustration of TM/γ-U alloy cluster model: the central U atom (1) replaced with TM atom [Kurihara et al., (2004)]

### 2.3 Theoretical x-ray photoelectron spectra

XPS (X-ray photoelectron spectroscopy) intensity was estimated using the following equation (Gellius, U., 1974),

$$I_l = \sum_i \sigma_i P_{il} . \tag{4}$$

Here, $I_l$ is the probability of photo-ionization from the $l$th MO level, $\sigma_i$ is the photo-ionization cross-section of the $i$th atomic orbital, and $P_{il}$ is the population of the $i$th atomic orbital in the $l$th MO. The photo-ionization cross-section of individual AOs was taken from the data estimated by *ab initio* calculations (Scofield, J. H., 1976). The cross-section of the U7p was ignored in the present calculations, because it was too small to contribute to the XPS intensity. For evaluating $P_{il}$, the gross population of individual AOs for each MO was evaluated by Mulliken population analysis (Mulliken et al., 1955a, 1955b, 1955c, 1995d). Theoretical XPS spectra were obtained by replacing each stick peak with a Lorentzian curve with a full width at the half maximum (FWHM) of 1.46 eV. We then compared these with the experimental XPS spectra of UC solid (Ejima, T., et al., 1993).

In a similar manner to that for UC solid, theoretical XPS spectra of α-U solid were obtained by replacing each stick peak with that with a FWHM of 1.6 eV. Then, we compared these with the experimental XPS for α-U metal (Fuggle, J. C., et al., 1982).

## 3. Relativistic effects in molecules

In this section, we briefly explain relativistic effects on the electronic structure and chemical binding in molecules for understanding the results of uranium compounds introduced in Sections 4 and 5. Prior to describing the effects in molecules, we begin to explain the relativistic effects in atoms. The Bohr radius ($a$) and the energy ($E$) of hydrogen-like atoms are respectively given by the following equations,

$$a = \frac{n^2\hbar^2}{mZe^2}, \tag{5}$$

and

$$E = -\frac{me^2Z^2}{2\hbar^2 n^2}. \tag{6}$$

Here, $m$ is the mass of electron, $n$ is the principal quantum number, $Z$ is the atomic number, $e$ is the elementary charge, $\hbar$ is Planck's constant. In addition, the ratio of the speed of 1s electron ($v_{1s}$) to that of light ($c$) can be written by Z/137. For U atom with Z=92, the $v_{1s}$ can be estimated to be 0.67$c$, which indicates that the relativistic effects should be considered in quantum formula. Even for the U6s valence atomic orbital, the $v_{6s}$ is calculated to be 0.1$c$, implying that the relativistic effects can appear in the valence region that is related to chemical bonding. According to the theory of relativity, the mass of electron is influenced by the relativistic effects, which can be given by the following equation,

$$m = \frac{m_0}{\sqrt{1-\left(\frac{v}{c}\right)^2}}. \tag{7}$$

Here, $m_0$ and $v$ are the rest mass and speed of electron, respectively. Accordingly, one can see from Eqs. (5) and (6) that when $m$ increases with increasing Z, Bohr radius and energy become contracted and increased, respectively. A heavier element exhibits more remarkable change in both $a$ and $E$ by the relativistic effects.

The relativistic effects on AOs can be classified into two categories: direct and indirect effects. The former effects result in the relativistic contraction of inner-shell orbitals (e.g. s and p orbitals). In addition, since the total angular momentum $j$ (= $l$ + $s$) is the quantum number in Dirac's equation, the energy splitting due to the spin-orbit interactions is simultaneously taken into account (e.g. 6p splits to $6p_{3/2}$ and $6p_{1/2}$) in the equation. On the other hand, the indirect effects cause the expansion of outer AOs (e.g. d and f AOs) due to the screening of nuclear charges by the contraction of the inner-shell AOs. As described above, these relativistic effects become magnified with increasing Z, and remarkably affect both valence electronic structure and chemical bonding of condensed matters (molecules, complexes, and solids) containing heavy elements such as uranium. We next explain how the relativistic effects affect the valence electronic structure and chemical bonding of $UF_6$ as an example.

### 3.1 Relativistic effects on valence electronic structures

Figure 5 shows the comparison between non-relativistic and relativistic one-electron energies of valence MOs for the ground state of $UF_6$ that is often used as a material for uranium enrichment. Here, MOs were classified into two parts based on inversion symmetry: "gerade" and "ungerade" shown on the left and right sides, respectively.

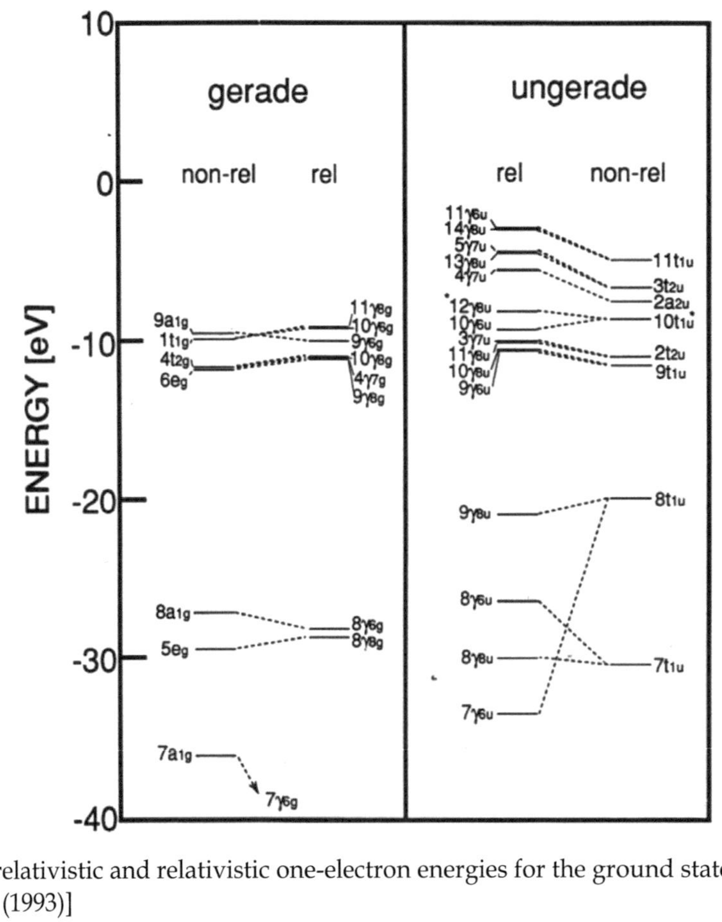


Fig. 5. Non-relativistic and relativistic one-electron energies for the ground state of $UF_6$ [Onoe et al., (1993)]

As shown in Fig. 5, two kinds of relativistic effects appear in the valence electronic structure: the energy splitting for several MOs and the upward or downward shift in the one-electron energy for individual MOs. In particular, since the energy splitting ($\Delta E$) between the U6$p_{1/2}$ and U6$p_{3/2}$ ($\Delta E$ = 8.9 eV) is larger than that for both U5f ($\Delta E$ = 0.8 eV) and U6d ($\Delta E$ = 0.5 eV) AOs, the non-relativistic 7$t_{1u}$ and 8$t_{1u}$ MOs caused by the U6p-F2s interactions exhibit the noticeable splitting to form the four corresponding relativistic MOs. Indeed, we succeeded in assigning these four relativistic MOs in the valence photoelectron spectra of $UF_6$ (Onoe et al., 1994b).

### 3.2 Relativistic effects on chemical bonding

To understand how the relativistic effects influence the chemical bonding of $UF_6$, we have investigated the contribution of individual AOs to the chemical bonding by examining the non-relativistic and relativistic radial wave functions [Onoe et al., 1993]. Figure 6 shows the radial wave functions of the uranium valence atomic orbitals (5f, 6s, 6p, 6d) obtained using non-relativistic and relativistic density-functional calculations. For the relativistic radial wave functions, the small components were omitted and only the large ones were shown in Fig. 6, because the former ones play a minor role of contributing to chemical bonding. Figure 6 suggests that the relativistic contraction and expansion of the valence AOs at the U-F bond length significantly affect the strength of chemical bonding of $UF_6$.

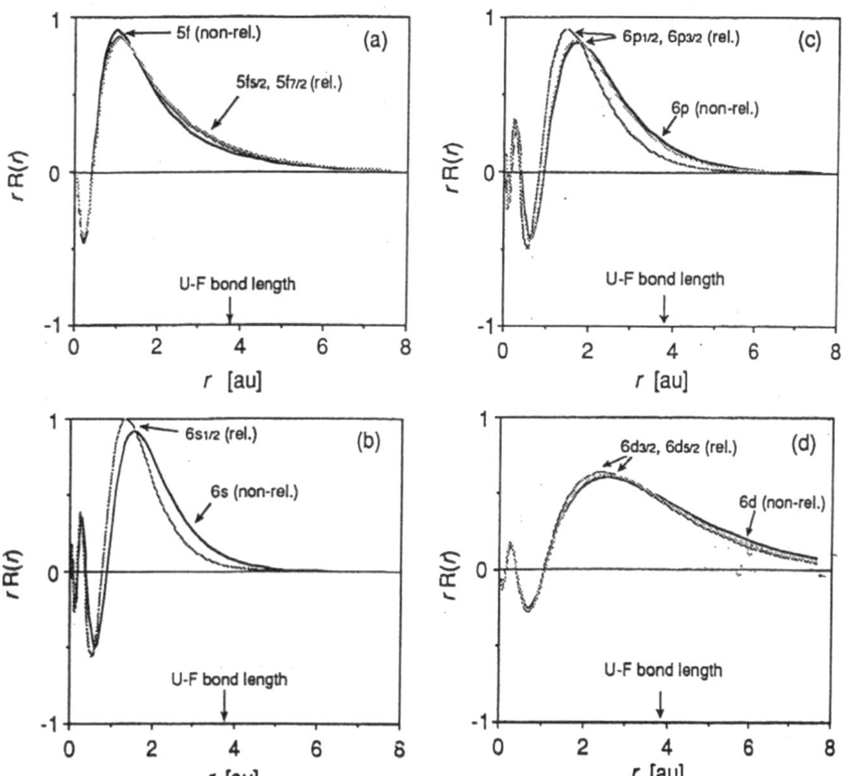


Fig. 6. Non-relativistic and relativistic radial wave functions [$R(r)$] of U valence atomic orbitals: (a) U5*f*, (b) U6*s*, (c) U6*p*, and U6*d* [Onoe et al., (1993)]

Since the orbital overlap population (OOP) is a good indicator for the strength of chemical bonding between AOs, we examined the OOP between individual U and F valence AOs, along with the bond overlap population (Bo) between U and F atoms, in order to clarify the contribution of the relativistic contracted and expanded AOs to the chemical bonding of $UF_6$.

Table 1 summarizes the OOP and Bo obtained from Mulliken population analysis for the non-relativistic and relativistic calculations. Here, the positive and negative signs of OOP imply the bonding and anti-bonding interactions, respectively. For an example, the strength of anti-bonding U6s-F2s and U6s-F2p interactions becomes weakened from non-relativistic to relativistic calculations, because the relativistic contraction of the U6s radial wave function reduces the U6s-F2s and U6s-F2p OOPs in the U-F bond region, as shown in Fig. 6. This results in the strengthening of the U-F bond. On the other hand, the relativistic expansion of the U5f and U6d radial wave function in the U-F bond region strengthens the U5f-F2p and U6d-F2p bonding interactions. The relativistic changes in the U valence AOs cause a large difference in the U-F chemical bond, as indicated in Table 1.

| | | DV-HFS (nonrelativistic) F | | DV-DS (relativistic) F | |
|---|---|---|---|---|---|
| | | 2*s* | 2*p* | 2*s* | 2*p* |
| U | 5*f* | 0.07 | 1.07 | 0.08 | 1.17 |
| | 6*s* | −0.16 | −0.54 | −0.05 | −0.29 |
| | 6*p* | −0.56 | −0.71 | −0.42 | −0.85 |
| | 6*d* | 0.18 | 1.54 | 0.16 | 1.77 |
| Total U–F bond overlap population | | +1.21 | | +2.15 | |

Table 1. The overlap populations between U and F valence AOs for non-relativistic and relativistic calculations, along with the U-F bond overlap population [Onoe et al., (1993)]

## 4. Application to assign the valence X-ray photoelectron spectra of uranium metal and compounds

### 4.1 Uranium carbides

Figure 7 shows the experimental valence XPS spectra of UC (Ejima, et al., 1993) (a) and theoretical spectra for the $UC_6$ (b), $CU_6$ (c), and $CU_6C_{18}$ (d) cluster models shown in Fig. 2, along with the partial density-of-states (pDOS) of the U5f and U6d AOs (e) and of the C2s and C2p AOs (f). As shown in Fig. 7(a), UC has the intense peak "1" at the Fermi level ($E_F$), the shoulder peak "2" at around -2 eV, and the broad weak peak "3" at around -10 eV. We next compared theoretical spectra [Fig. 7(b)-(d)] with the experimental result. As shown in Fig. 7(b), the $UC_6$ cluster model well reproduced both relative intensity and position (binding energy) for the peaks "1" and "3", but the shoulder peak "2" unfortunately seemed to be reproduced. On the other hand, the $CU_6$ cluster model showed a peak corresponding to the peak "2" besides the peaks "1" and "3", but it was completely separated (not a shoulder peak) from the peak "1". Accordingly, it is found that the $CU_6$ cluster model reproduces the experimental spectra better than the $UC_6$ cluster model. When the $CU_6$

cluster is surrounded with eighteen C atoms, a $CU_6C_{18}$ embedded cluster model is expected to reproduce the experimental spectra better than the $CU_6$ cluster. Comparison between Figs. 7(a) and 7(d) indicates that the theoretical spectra well reproduce the whole experimental one.

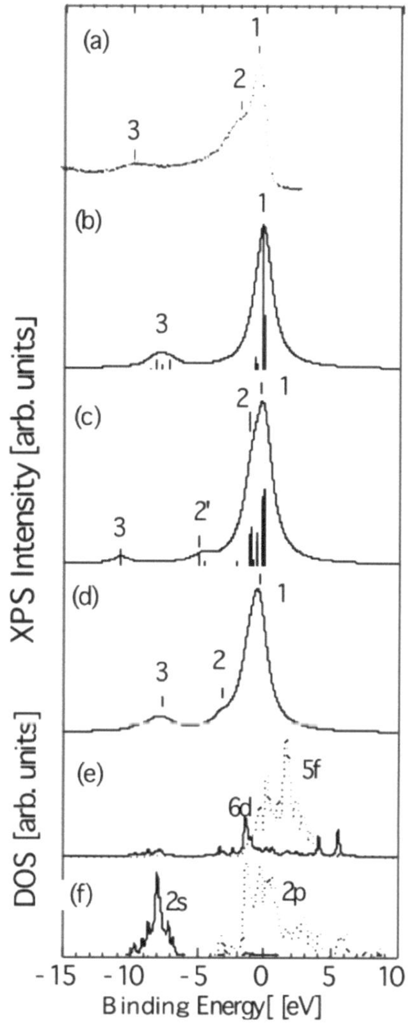


Fig. 7. Theoretical and experimental valence x-ray photoelectron spectra for UC: (a) experiment [Ejima. T., et al., (1993)] and theoretical spectra obtained using (b) $UC_6$, (c) $CU_6$, and (d) $CU_6C_{18}$ cluster models, along with partial density-of-states (pDOS) for (e) the U5f (dashed line) and U6d (solid line) AOs and (f) the C2s (solid line) and C2p (dashed line) AOs [Kurihara et al.,(1999)]

To assign the three peaks shown in Fig. 7(a), we next examined the pDOS based on the results of the $CU_6C_{18}$ embedded cluster model. Figures 7(e) shows the pDOS of the U5f (dashed line) and U6d (solid line) AOs, whereas Figure 7(f) shows that of the C2s (solid line) and C2p (dashed line) AOs. Comparison between Figs. 7(a) and 7(e, f) indicates that the intense peak “1” is mainly attributed to the U5f AO which has a large photo-ionization cross-section and the shoulder peak "2" is attributed to the U6d-C2p AOs, whereas the broad weak peak "3" is mainly attributed to the C2s AO.

In previous reports by the other groups, Schalder et al reported that the intense peak "1" was due to the U5f-U6d bands, while the shoulder "2" due to the U6d-C2p band (Schadler, G. H., 1990). On the other hand, Ejima et al concluded that a small amount of the U5d component contributes to the shoulder "2". However, because the $U5d_{5/2}$ and $U5d_{3/2}$ AOs are respectively located at -91 eV and -99 eV, the U5d components can be considered to play a minor role for the shoulder "2". In fact, the present calculations show that the U5d components have no contribution to the valence electronic structure of UC.

### 4.2 α-Uranium metal

Figure 8 shows (a) the experimental (Fuggle, J. C., et al., 1974) and theoretical x-ray photoelectron spectra obtained using the $U_9$ cluster model (Kurihara, M., et al., 2000) for α-U metal, (b) the pDOS of U valence AOs, and (c) the magnified pDOS in the vicinity of $E_F$. As shown in Fig. 8(a), it is interesting to note that the theoretical spectra well reproduced the experimental one of α-U metal, in spite of using the minimum $U_9$ cluster model shown in Fig. 8. By comparison between Figs. 8(a) and 8(b), it is clearly found that the peaks "B" and "C" are only attributed to the $U6p_{3/2}$ and $U6p_{1/2}$ AOs, respectively. In a similar manner, the

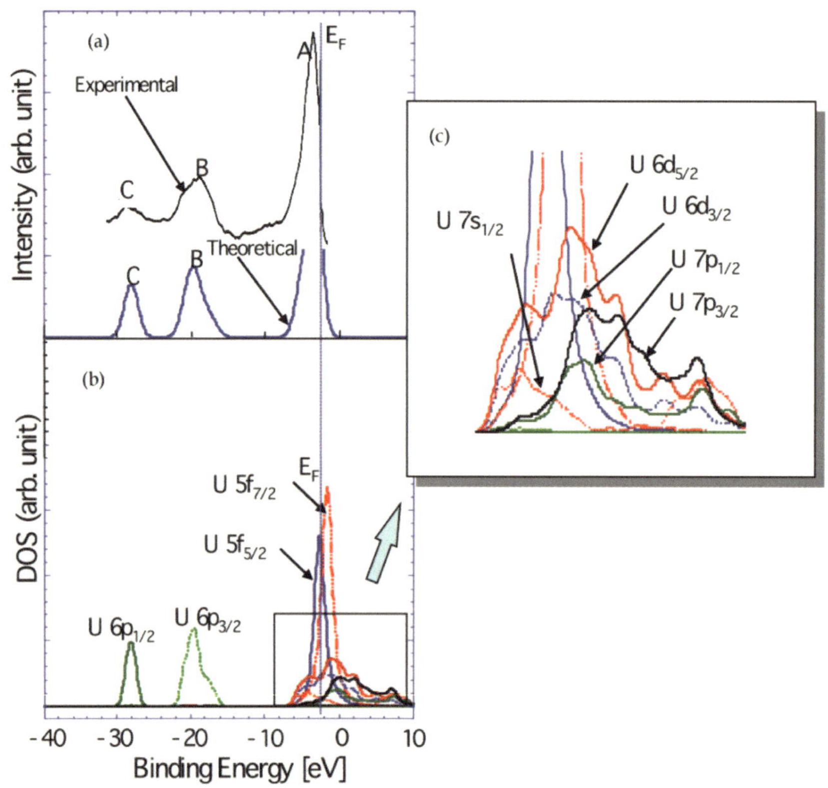


Fig. 8. (a) Experimental [Fuggle, et al., (1974)] and theoretical spectra, (b) partial density-of-states (pDOS) of U valence AOs and (c) magnified pDOS in the vicinity of $E_F$ [Kurihara et al., (2000)]

intense peak "A" appearing at $E_F$ is mainly attributed to the U5$f_{7/2}$ and U5$f_{5/2}$ AOs. As shown in Fig. 8(c), since the other valence AOs such as the U6d, U7s, and U7p are located around $E_F$, they contribute to the peak "A" to some extent. However, because their photo-ionization cross-sections are much smaller than that of the U5f AOs, they are considered to have minor contributions.

In previous band calculations (Yamagami. H., & Hasegawa. A., 1990), they compared the DOS structures with the experimental spectra, and discussed the assignment of each peak qualitatively. On the contrary, the present method can calculate XPS intensity of individual MOs for the cluster model used as the minimum unit of α-U metal. Thus we have quantitatively obtained the valence XPS spectra, together with pDOS of individual AOs contained in each MOs. This enables us to assign individual peaks satisfactorily.

## 5. γ-Uranium/transition metal alloys

As introduced in the previous section, the present R-DFT method well reproduced the experimental photoelectron spectra of UC and α-U solid and assigned the origins of individual peaks. In this section, we applied this method to understand what electronic factors play significant roles of the alloying behavior for γ-U/TM metal alloys that are one of the candidates as nuclear fuels for advanced reactors.

### 5.1 Correlation between the Md (or effective charges) and the maximum solid solubility

The Md has been found to play an important role of alloying for $Ni_3Al$ (Morinaga et al., 1984) and bcc Fe ((Morinaga et al., 1985a). In this section, we first present the correlation between the maximum solid solubility (MSS) and Md for γ-U/TM alloys (Kurihara et al., 2004, 2008).

Figure 9 shows the plot of MSS as a function of Md, where the regression lines were obtained using a least square method. It is found that MSS is exponentially dependent on Md except for γ-U/Ta and γ-U/W alloys, though the reason of this exception is not still clearly understood. This exponential dependence will be discussed from a thermodynamic standpoint in Section 5.4.

Since Md is related to the charge transfer (CT) between the TM and γ-U, we next examined the correlation between MSS and CT (Kurihara et al., 2008). Figure 10 shows the plot of Md as a function of atomic number (Z) for individual TMs in γ-U/TM alloys, along with the U6d energy of U(1) prior to TM substitution. The sign and amount of the difference between Md and U6d energies determine the direction and amount of CT between TM and γ-U, respectively. The CT takes place from TMs to γ-U when Md level is higher than the U6d one, whereas CT from γ-U to TMs takes place when Md level is lower than the U6d one. In addition, a larger difference between Md and U6d energies results in a larger amount of CT between them. From the results of Figs. 9 and 10, a smaller difference between the Md and U6d levels results in a larger MSS, as previously reported by Morinaga et al. (Morinaga et al., 1984, 1985a,b). Namely, a smaller amount of CT between TM and γ-U provides a larger MSS for γ-U/TM alloys.

Since the amount of CT is directly related to the effective charge, we next examined the correlation between MSS and the effective charges on TM and U atoms (Kurihara et al.,

2008). Figure 11 shows that MSS exhibits a negatively exponential relationship with respect to the effective charge. This implies that MSS exhibits a negatively exponential dependence on the amount of CT between the TM and U atoms. Furthermore, because the effective charge indicates the degree of ionic bonding interaction between TMs and γ-U, it can be said that MSS shows a negatively exponential dependence on the strength of ionic bonding between TM and γ-U atoms. This indicates that a large ionic bonding between TM and γ-U atoms results in a smaller MSS for γ-U/TM alloys.

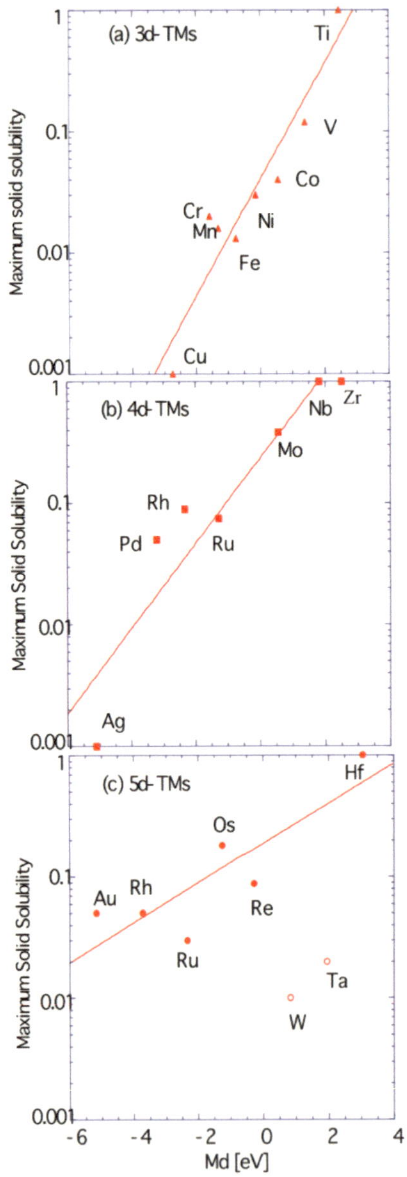


Fig. 9. Plot of the maximum solid solubility (MSS) as a function of Md for γ-U/TM alloys [Kurihara et al., (2008)]

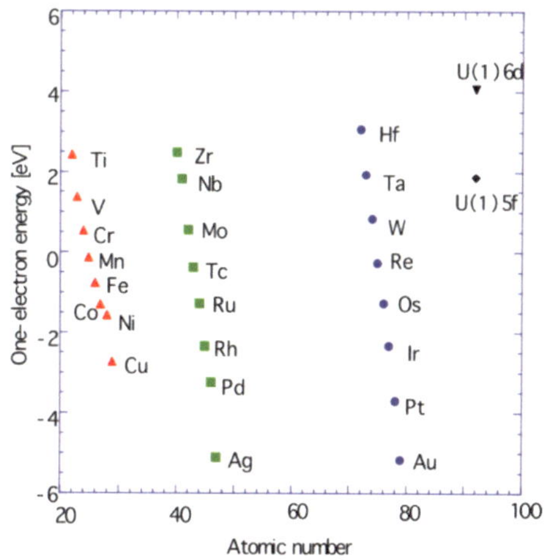


Fig. 10. Plot of Md as a function of atomic number (Z) for individual TMs in γ-U/TM alloys, along with the U6d and U5f energies of U(1) prior to TM substitution [Kurihara et al., (2008)]

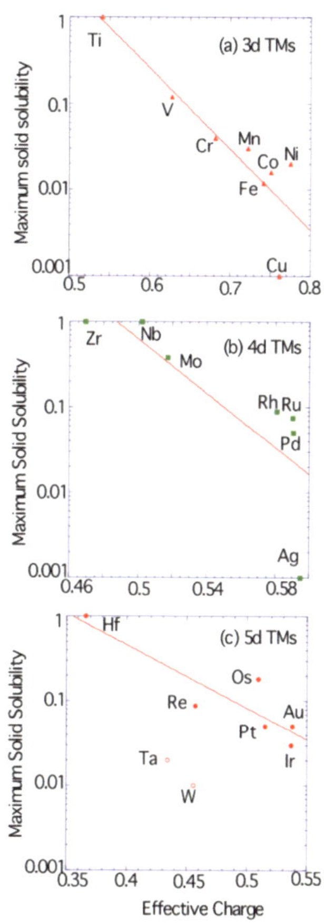


Fig. 11. The correlation between the maximum solid solubility (MSS) and the charge transfer (CT) between TM and U atoms for γ-U/TM alloys [Kurihara et al., (2008)]

Since the bonding between TMs and γ-U contain not only ionic but also covalent interactions, we next discuss the correlation between MSS and the covalent interactions between TMs and γ-U.

### 5.2 Correlation between orbital overlap populations and the maximum solid solubility

We examined the correlation between TMd-U6d OOP and MSS for γ-U/TM alloys, because OOP is a good indicator for the strength of the covalent bonding between TM and U atoms. Figure 12 shows the plot of MSS as a function of OOP between TMd and U6d AOs

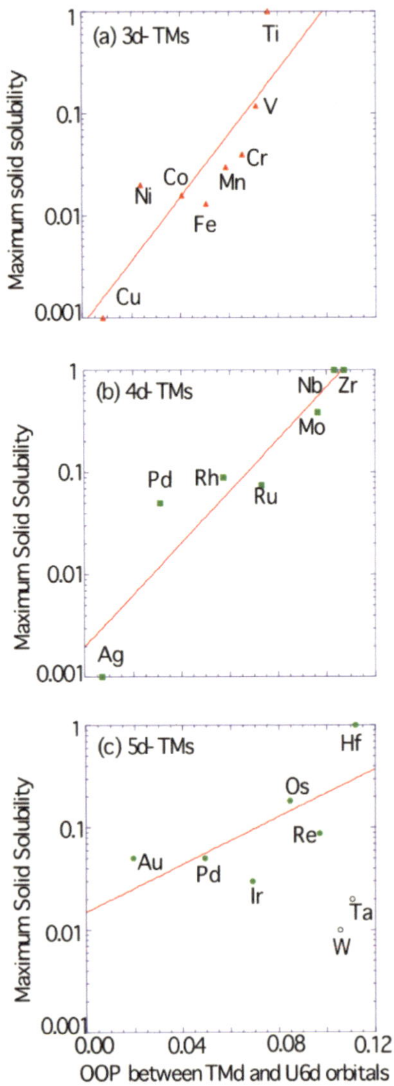


Fig. 12. The correlation between the maximum solid solubility (MSS) and the orbital overlap population (OOP) between TMd and U6d atomic orbitals: (a) 3d TMs, (b) 4d TMs and (c) 5d TMs [Kurihara et al., (2008)]

(Kurihara et al., 2008), and demonstrates that MSS exhibits an exponential dependence on the strength of the U6d-TMd covalent bonding. Namely, a stronger covalent bonding between TM and U atoms provides a larger MSS for TMs alloyed into $\gamma$-U. This is opposite to the correlation between MSS and the TM-U ionic bonding.

We will discuss the reason why MSS is exponentially proportional to OOP as well as Md in Section 5.4.

### 5.3 Role of U6d-TMd orbital interactions in alloying of $\gamma$-U/TMs

Figure 13 shows the plot of MSS as a function of the energy difference ($\alpha_U$ - $\alpha_{TM}$) between the U6d and TMd AOs (Kurihara et al., 2011). The energy difference ($\alpha_U$ - $\alpha_{TM}$), which is often used to discuss the CT between U and TM atoms, shows that MSS exhibits an inversely exponential dependence on ($\alpha_U$ - $\alpha_{TM}$), thus a smaller value of ($\alpha_U$ - $\alpha_{TM}$) provides a larger MSS. Since the energy difference ($\alpha_U$ - $\alpha_{TM}$) is related to the magnitude of the CT between TM and $\gamma$-U, Figure 13 implies that a smaller CT between them gives rise to a larger MSS for $\gamma$-U/TM alloys.

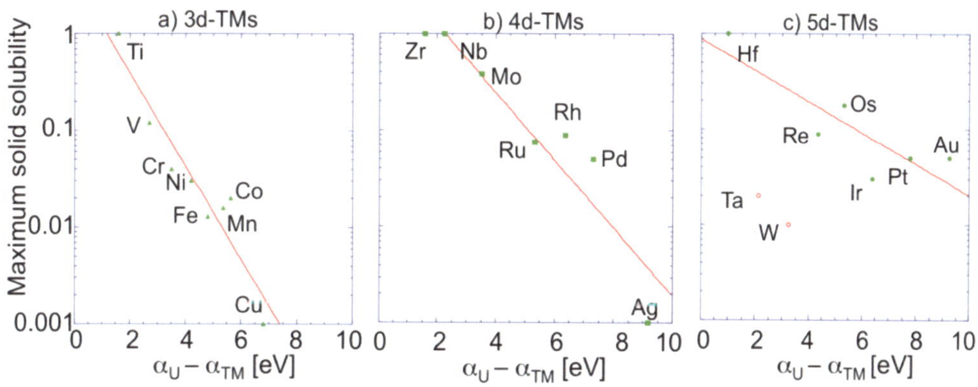


Fig. 13. Plot of the maximum solid solubility (MSS) as a function of the energy difference between the U6d ($\alpha_U$) and TMd ($\alpha_{TM}$) AOs [Kurihara et al., (2011)]

Figure 14 shows the correlation between the TMd-U6d OOP and MSS for $\gamma$-U/TM alloys (Kurihara et al., 2011), and demonstrates the exponential dependence of MSS on the OOP for all $\gamma$-U/TM alloys except for Ta and W elements. Thus, an increase in the OOP results in an increase in MSS of $\gamma$-U/TM alloys.

From the results of Figs. 13 and 14, one can see that the interactions between TMd and U6d AOs play an important role of determining the magnitude of MSS for $\gamma$-U/TM alloys. In other words, the TMd-U6d orbital interactions become a key parameter for estimating MSS of $\gamma$-U/TM alloys. We next discuss the physical meaning of the TMd-U6d orbital interactions in the magnitude of MSS.

Figure 15 schematically illustrates the stabilization energy ($\Delta E$) caused by the U6d-TMd orbital interactions from a viewpoint of molecular orbital theory (top) and the correlation between MSS and $\Delta E$ or the U6d-TMd energy difference (bottom). In the framework of the simple Hückel approximation (Hückel, E, 1931), $\Delta E$ is obtained as,

$$\Delta E = \frac{1}{2}\left[\sqrt{(\alpha_U - \alpha_{TM})^2 + 4\beta^2} - (\alpha_U - \alpha_{TM})\right]. \tag{8}$$

Here, $\alpha_U$ and $\alpha_{TM}$ respectively denote the U6d and TMd energies, and $\alpha_U$ is equal to or greater than $\alpha_{TM}$ ($\alpha_U \geq \alpha_{TM}$). $\beta$ denotes the resonance integral between the TMd and U6d AOs, which can be written as,

$$\beta = \frac{1}{2}(\alpha_U + \alpha_{TM})KS$$

Where, $K$ is a constant, and $S$ is the overlap integral between the TMd and U6d AOs. Namely, $\beta$ is proportional to $S$, thus proportional to the TMd-U6d OOP.

To discuss the correlation between $\Delta E$ and Md/OOP more clearly, we introduce the following two variables, $t = (\alpha_U - \alpha_{TM}) \big/ 2|\beta|$ and $F(t) = \sqrt{t^2 - 1} - t$, into Eq. (8). Then, $\Delta E$ can be rewritten as,

$$\Delta E = F(t) \cdot |\beta|. \tag{9}$$

By considering the range of the two quantities on the right-hand side of Eq. (7) (i.e., $t \geqq 0$ and $0 < F(t) \leqq 1$), we obtained the range of the stabilization energy,

$$0 < \Delta E \leqq |\beta|. \tag{10}$$

Accordingly, $\Delta E$ is the maximum ($= |\beta|$) when $F(t)$ is unity at $t = 0$, that is, $\alpha_U = \alpha_{TM}$. On the contrary, $\Delta E$ becomes the minimum (near equal to zero) when ($\alpha_U$ - $\alpha_{TM}$) is much larger than unity (≫1). Consequently, a larger MSS for γ-U/TM alloys is obtained at a lower value of ($\alpha_U$ - $\alpha_{TM}$) (corresponding to a larger $\Delta E$), whereas a small MSS is obtained at a larger value of ($\alpha_U$ - $\alpha_{TM}$) (corresponding to a smaller $\Delta E$). Consequently, it can be concluded that the magnitude of $\Delta E$ caused by the TMd-U6d orbital interactions plays a key role of determining the magnitude of MSS.

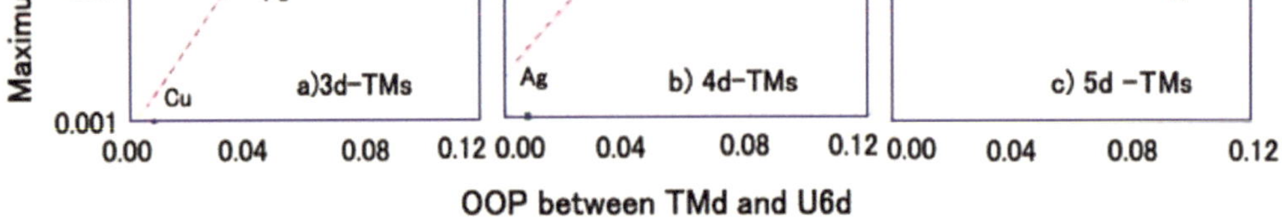


Fig. 14. Plot of the maximum solid solubility (MSS) as a function of OOP between the U6d and TMd AOs for γ-U/TM alloys [Kurihara et al., (2011)]

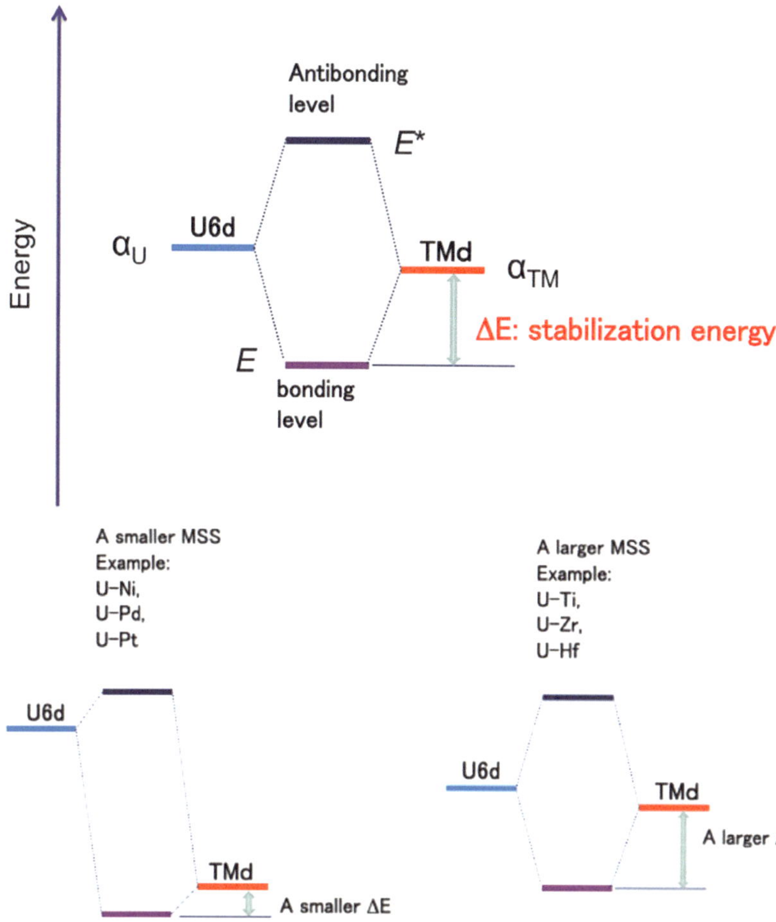


Fig. 15. Schematic illustration of the TMd-U6d orbital interactions based on molecular orbital theory (Top) and of the correlation between MSS and the stabilization energy ($\Delta E$) caused by their orbital interactions (Bottom) [Kurihara et al., (2011)]

### 5.4 Exponential dependence of MSS on Md and OOP

As shown in Figs. 9 and 12, MSS exhibits an exponential dependence on both Md [in other words, the U6d-TMd energy difference ($\alpha_U$ - $\alpha_{TM}$)] and the U6d-TMd OOP. In the present work, we employed the cluster model (see Fig. 4) for γ-U/TM alloys by substituting the central U atom with TM atoms. The following equilibrium has been considered based on the cluster model,

$$U_9 + TM \Leftrightarrow U_8TM + U \,. \tag{11}$$

Here, TM denotes a TM atom in the TM bulk phase, while $U_9$/$U_8TM$ and U respectively denote a cluster and a U atom in the γ-U bulk phase. As the equilibrium constant ($K$) in Eq. (11) increases, the concentration of $U_8TM$ increases, thus increasing MSS. On the other hand, as $K$ decreases, the $U_8TM$ concentration correspondingly decreases, thus reducing MSS.

Accordingly, MSS is proportional to *K*. Then the relation between MSS and *K* can be written as,

$$MSS \propto K = Ae^{-\Delta G/RT} \quad (12)$$

Where, A is a constant, $\Delta G$ denotes the difference in Gibbs free energy before and after the substitution, *R* is the gas constant, and *T* is the absolute temperature. Since the lattice relaxation associated with substitution of TM into γ-U solid was negligible in the present calculations, the entropy remained constant ($\Delta S = 0$) before and after the TM substitution. Thus, $\Delta G$ is equal to the enthalpy difference ($\Delta H$) before and after the substitution. Accordingly, Eq. (12) can be rewritten as,

$$MSS \propto K = Ae^{-\Delta G/RT} = Ae^{-\Delta H/RT} . \quad (13)$$

Since $\Delta H$ includes $\Delta E$ caused by the TMd-U6d orbital interactions, the following relationship is obtained,

$$MSS \propto K = Ae^{-\Delta G/RT} = Ae^{-\Delta H/RT} \propto e^{\Delta E/RT} . \quad (14)$$

Since $\Delta E$ is proportional to both ($\alpha_U - \alpha_{TM}$) and *S*, it can be understood that MSS depends exponentially on both Md and OOP, as shown in Figs. 9 and 12.

In the above discussion, Md and OOP can be regarded as a good parameter to determine the magnitude of MSS. Then we next made the OOP–Md map for γ-U/TM alloys (Kurihara et al, 2011), as shown in Fig. 16. Interestingly, γ-U/TM alloys with a smaller MSS appear in a

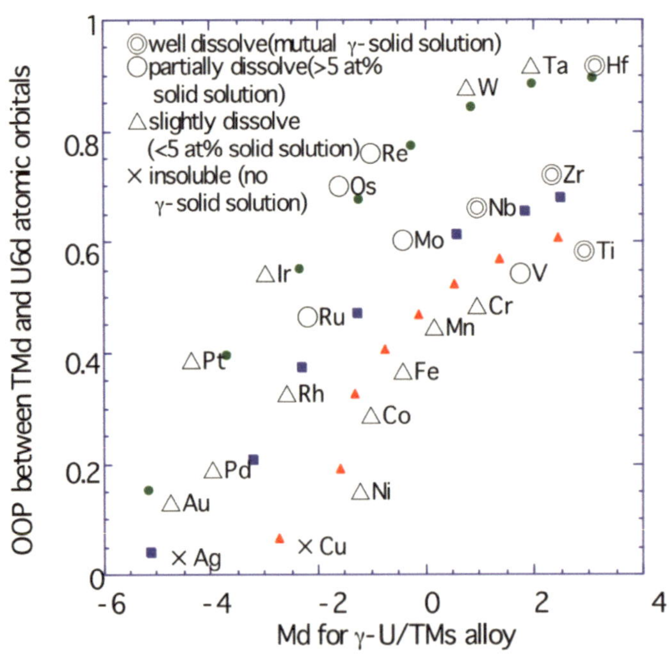


Fig. 16. OOP-Md plot of γ–U/TM alloys [Kurihara et al., (2011)]

lower left hand on the map, whereas those with a higher MSS in an upper right hand. Consequently, the OOP–Md map could be useful for designing γ-U/TM alloys. We believe that the present findings can be used for not only the γ-U/TM alloy system but also other actinide alloys used as nuclear fuels for advanced reactors.

## 6. Summary and perspective

We have performed R-DFT calculations for understanding and designing γ-U/TM alloys as a typical example of nuclear fuels. Md and OOP have a good correlation with MSS of γ-U/TM alloys, and become a more suitable parameter to determine the magnitude of MSS than metallic radius as previously proposed by Hume-Rothery. The present parameters, Md and OOP, demonstrated that a stronger covalent bonding between TMd and U atoms gives rise to a larger MSS of γ-U/TM alloys, whereas a stronger ionic bonding between them gives rise to a smaller MSS.

In summary, the magnitude of MSS was successfully explained in terms of the stabilization energy ($\Delta E$) caused by the U6d-TMd orbital interactions for γ-U/TM alloys. In addition, the exponential dependence of MSS on Md and OOP was also interpreted by considering the equilibrium constant based on thermodynamics of the substitution model for γ-U/TM alloys. The OOP-Md map was useful to determine the magnitude of MSS for γ-U/TM alloys. Finally, the present approach could be applied to understand and design the other alloy systems used as nuclear fuels for advanced reactors.

## 7. Acknowledgments

The authors are great thankful to Dr. M. Hirata (Japan Atomic Energy Agency) for valuable and helpful comments on this article, and also thankful to Profs. H. Adachi and T. Mukoyama (Kyoto University), and Prof. R. Sekine (Shizuoka University) for fruitful discussion on uranium molecules such as $UF_6$. Finally, this review article is sincerely dedicated to late Prof. H. Nakamatsu (Kyoto University).

## 8. References

Adachi. H., et al., (1978). *J. Phys. Soc. Jpn.* Vol. 45, pp. 875-883.
Alonso. P. R., et al., (2009). *Phys.* Vol. B 404, pp. 2851-2853.
Bethe, H., (1929). *Ann. Phys.* Vol. 3, pp. 133-208.
Buzzard. R. W., (1955). Progress Report-*Alloying Theory*, NBS-4032.
Chernock. W. & Horton. K. E., et al., (1994). IAEA *TECDOC-791*, p. 68, Vienna.
Chiotti. P., et al., (1981). IAEA *STI/PUB/424/5*, Vienna.
Ejima, T., et al., (1993). *Physica B*, Vol. 186/188, pp 77-79.
Erode, P., et al., (1983). *The Physics of Actinides Compounds,* Plenum Press. New York.
Fuggle, A. F., et al., (1974). *J. Phys. F,* Vol. 4, pp 335-342.
Gelius, U., (1974). *J. Electron Spectrosc. Relat. Phenom.* Vol. 5, pp. 985-1057.
Hirata. M., et al., (1997). *J. Electron Spectrosc. Relat. Phenom.* Vol. 83, pp. 59-64.
Holden. A. M., (1958). *Physical Metallurgy of Uranium*, Addison-Wesley, New York.
Hückel . E., (1931). *Z. F. Phys.* Vol. 70, pp. 204-286.

Hume-Rothery. M. & Raynor. G. W., (1954). *Structure of Metals and Alloys,* Institute of Metals, London.
Kaufman. A. R., (1962). *Nuclear Reactor Fuel Elements,* Interscience/Butterworth, USA.
Kim. K. –H., et al., (1999). *J. Nucl. Mater.* Vol. 270, pp. 315-321.
Kim. K. -H., et al., (2002). *Nucl. Eng. Design* Vol. 211, pp. 229-235.
Kurihara. M., et al., (1999). *J. Alloys Compd.* Vol. 283, pp. 128-132.
Kurihara. M., et al., (2000). *J. Nucl. Mater.* Vol. 281, pp. 140-145.
Kurihara. M., et al., (2004). *J. Nucl. Mater.* Vol. 326, pp. 75-79.
Kurihara. M., et al., (2008). *Prog. Nucl. Ener.* Vol. 50, pp. 549-555.
Kurihara. M., et al., (2011). *J. Alloys Compd.* Vol. 509, pp.1152-1156.
Landa. A., et al., (2009). *J. Alloys Compd.* Vol. 478, pp. 103-110.
Li. Z. S., et al., (2009). *J. Alloys Compd.* Vol. 476, pp. 193-198.
Meyer, J., et al., (1989). *Comput. Phys. Commum.* Vol. 54, pp. 55-73.
Meyer. M. K., et al., (2002). *J. Nucl. Mater.* Vol. 304, pp. 221-236.
Morinaga. M., et al., (1984). *J. Phys. Soc. Jpn.* Vol. 53, pp. 653-663.
Morinaga. M., et al., (1985a). *J. Phys. F, Met. Phys.* Vol. 15, pp. 1071-1084.
Morinaga. M., et al., (1985b). *Philos. Mag. A* Vol. 51, pp. 223-246.
Morinaga. M., et al., (1985c). *Philos. Mag. A* Vol. 51, pp. 247-252.
Morinaga. M., et al., (1991). *J. Phys.: Condens. Matter.* Vol. 3, pp. 6817-6827.
Morinaga. M., et al., (2005). *Alloy design based on the DV-Xα cluster method.* (Adachi, H et al., Eds.), Hartree-Fock-Slater Method for Material Science, The DV-Xα Method for Design and Characterization of Materials, Springer Series in Materials Science, Vol. 84, pp. 23-48, ISBN-10-3-540-24508-1, Springer, Berlin, Heidelberg, New York.
Mulliken. R. S., (1955a). *J. Chem. Phys.* Vol. 23, pp. 1833-1840.
Mulliken. R. S., (1955b). *J. Chem. Phys.* Vol. 23, pp. 1841-1845.
Mulliken. R. S., (1955d). *J. Chem. Phys.* Vol. 23, pp. 2343-2347.
Mulliken. R. S.,(1955c). *J. Chem. Phys.* Vol. 23, pp. 2338-2342.
Ogawa. T., et al., (1955). *J. Nucl. Mater.* Vol. 223, pp. 67-71.
Onoe. J., et al., (1992). *J. Electron Spectrosc. Relat. Phenom.* Vol. 60, pp. 29-36.
Onoe. J., et al., (1993). *J. Chem. Phys.* Vol. 99, pp. 6810-6817.
Onoe. J., et al., (1994). *J. Electron Spectrosc. Relat. Phenom.* Vol. 70, pp. 89-93.
Park. J. J. & Buzzard. R. W., (1957). *TID-7526 (Pt. 1)* pp. 69-102.
Pauling L., (1960). *The Nature of the Chemical Bond,* third ed. Cornell Univ. Press, New York.
Rosen. A & Ellis. D. E., et al., (1975). *J. Chem. Phys.* Vol. 62, pp. 3039-3049.
Rosen. A., et al., (1976). *J. Chem. Phys.* Vol. 65, pp. 3629-3634.
Schadler, G.H., (1990). *Solid State Commum.* Vol. 74, pp. 1229-1231.
Scofield, J. H., (1976). *J. Electron Spectrosc. Relat. Phenom.* Vol. 8, pp. 129-137.
Sedmidudbsky. D., et al., (2010). *J. Nucl. Mater.* Vol. 397, pp. 1-7.
Yamagami, H. & Hasegawa, A. (1990). *J. Phys. Soc. Jpn.* Vol. 59, pp. 2426-2442.
Zachariasen W. H., (1973). *J. Inorg. Nucl. Chem.* Vol. 35, pp. 3487-3497.

# 5

# First Principles Simulations on Surface Properties and Oxidation of Nitride Nuclear Fuels

Yuri Zhukovskii[1], Dmitry Bocharov[2,3], Denis Gryaznov[1] and Eugene Kotomin[1]
[1]*Institute of Solid State Physics,*
[2]*Faculty of Computing,*
[3]*Faculty of Physics and Mathematics, University of Latvia, Riga*
*Latvia*

## 1. Introduction

Uranium mononitride (UN) is an advanced material for the non-oxide nuclear fuel considered as a promising candidate for the use in Generation-IV fast nuclear reactors to be in operation in the next 20-30 years [1, 2]. UN reveals several advantages over a traditional $UO_2$–type fuel (*e.g.*, higher thermal conductivity and metal density as well as high solubility in nitric acid in the case of fuel reprocessing [2]). However, one of important problems with actinide nitrides is their effective oxidation in oxygen-containing atmosphere which can affect nuclear fuel performance [3, 4]. Thus, it is important to understand the mechanism of the initial stage of UN oxidation and to find proper solutions, in order to improve in the future the fabrication process of this nuclear fuel.

In the present Chapter we acquire information on the atomic and electronic structure of both perfect and defective UN surfaces and discuss a mechanism of early stages of its surface oxidation. This Chapter is based on our own theoretical studies [5-10] as well as available theoretical and experimental literature.

## 2. Literature review

### 2.1 Experimental study of UN properties

Uranium mononitride is a metallic compound with low electrical resistivity ($1.6 \cdot 10^{-4}$ Ω·cm) [11], possessing *fcc* structure (space group $Fm\bar{3}m$, Fig. 1) over a wide temperature range [1]. The UN lattice constant is sensitive to carbon impurities [1] being insensitive to small oxygen impurities [12]. High melting temperature (~2780±25 K) [1], high fissile atom density (14.32 g/cm$^2$ vs 10.96 g/cm$^2$ for $UO_2$ [13]) and high thermal conductivity (13 W/mK) [14] make UN fuel a prospective material for nuclear reactors [1].

Fascinating and often enigmatic array of UN magnetic and electronic properties is induced by U(5*f*) electrons which are found to be intermediate between the highly localized 4*f* electrons of the lanthanides and the strongly delocalized *d* valence electrons in the transition metals [15]. UN was found to be antiferromagnetic at temperatures below the Neel temperature ($T_N$ ~ 53 K), which was detected in the heat capacity measurements [1]. Investigation of the magnetic structure of UN was performed in 60s by means of neutron diffraction [11]. The magnetic structure known as ordering of the first kind, where ferromagnetic sheets parallel to the (001) planes are antiferromagnetically coupled, was deducted [11]. The value of 0.75 $\mu_B$ found for the magnetic moment of U at $T<T_N$ appears to be surprisingly small (the lowest amongst the uranium monopnictides UX, where X = P, As, Sb) [11].

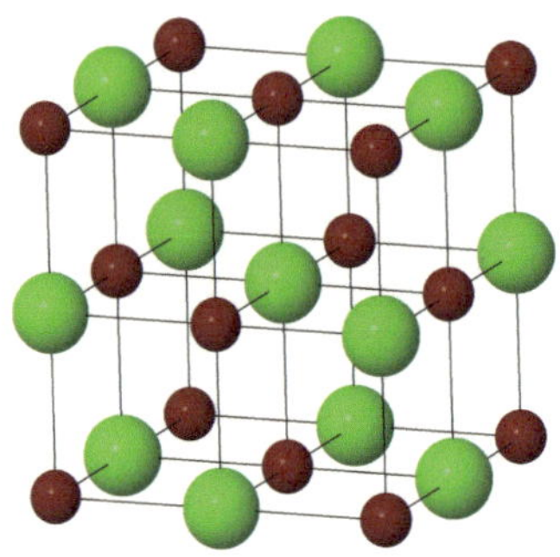


Fig. 1. An *fcc* structure of uranium mononitride.

Photoelectron spectroscopy also confirmed the complexity of UN. A very high density of states in proximity of the Fermi level was observed, which gives an evidence that the U(5*f*) electrons participate in bonding being strongly hybridized with the U(6*d*) electrons. The occupation of the conduction U(5*f*) band is 2.2 ± 0.5 *e*, of which ~1.8 *e* resides near the Fermi level [15]. In Ref. [16], the band structure of UN at 25 K was constructed taking into account the second derivative of high-resolution angle-resolved photoemission spectra. A highly dispersive band was observed for UN near the Fermi level centered at Γ(X) point, whose bottom is located at about 2 eV. First magneto-optical Kerr effect measurements on UN also showed narrow U(5*f*) band around the Fermi level as well as increased hybridization of the U(5*f*) states with U(6*d*) and N(2*p*) states as compared to similar data for heavier uranium monopnictides [17]. On the other hand, uranium nitride has the smallest U-U distance amongst the UX compounds (X=N, P, As, Sb, S, Se, and Te) which is equal to 3.46 Å being close to the critical 3.4 Å value given by Hill diagrams separating non-magnetic from magnetic compounds.

### 2.2 Interaction of uranium nitride with oxygen

The oxidation of uranium mononitride in an oxygen atmosphere was systematically studied in Ref. [18]. The two main types of UN samples were used in these experiments: powdered UN and smoothly polished UN pieces. Following a weight change of the UN powder sample during the oxidation process at elevated temperatures, a strong exothermic reaction was identified at 250°C characterized by rapid oxygen absorption. The weight was increased by 11.5%. The X-ray diffraction patterns of the intermediate product at temperatures 250-

260°C showed both weak diffraction lines corresponding to UN and very pronounced line broadening corresponding to $UO_2$. The polished UN pieces were used for kinetic study of UN oxidation. Measurements showed that the reaction rate is proportional to the area covered by the oxide or the oxidized volume. Analysis of both kinetic studies and X-ray diffraction data suggested that the isothermal oxidation of UN proceeds from the beginning of lateral spreading of the oxide, $UO_2(N)$, accompanied by a slight $N_2$ release and by the formation of $U_2N_3(O)$ during the reaction between UN and released nitrogen.

In Ref. [19], such characteristics as the chemical composition, phases, lattice parameter, sinterability, grain growth and thermal conductivity of the samples are investigated using chemical, X-ray and ceramographic analyses for pellets of uranium nitride powder containing certain amounts of oxygen (~0.3, ~0.6 and ~1.0 wt%) which are products of carbothermic reduction. Note, that conductivity of UN samples was found to be gradually decreasing under oxidation [20]. The principal results are that the average UN grain size of matrix phase decreases with increase of oxygen content. Moreover, thermal conductivity of the pellets containing about 1 wt% oxygen is lower than that of usual nitride pellets (containing 1000-2000 ppm oxygen) by 9-10% and 12-13% at 1000 and 1500 K, respectively.

In Ref. [21], direct ammonolysis of $UF_4$ was used, to synthesize $UN_2$ sample which was heated to 1100° C for 30 min inside the inert atmosphere producing these UN powder samples with $UO_2$ inclusions saturated at 5.0 wt%. The observed characteristic length distribution of particles ranges from 0.1 to 6 $\mu$m. The measured UN surface area was equal to 0.23 $m^2/g$. Both the electron microprobes and X-ray diffraction analysis showed that there are considerable amount of oxygen impurities in UN samples consisting of the primary UN phase and the secondary $UO_2$ impurity phase. This supports the conclusion that oxide impurities are likely to be formed by a diffusive process from the chemical environment and, thus, they are also likely to be formed along the particle surface. Concentration of oxygen impurities increases upon exposure to air: UN sample exposed for 3 months shows the growth of oxide contamination. The quantitative analysis performed for the XRD patterns showed that the $UO_2$ concentration increases from 5.0 wt% to 14.8 wt% over this time period [21].

The UPS measurements performed for thin layers of $UO_2$, UN, $UO_xN_y$ and $UO_xC_y$ using He-II 40.81 eV excitation radiation produced by a UV rare-gas discharge source were described in [20, 22]. These layers were prepared *in situ* by reactive DC sputtering in an Ar atmosphere. Fig. 2 shows that U(5*f*) states form a peak close to the Fermi level (0 eV), which proves their itinerant character. The valence band spectrum of $UO_xN_y$ shows a broad band interpreted as superposition of the narrow O(2*p*) and N(2*p*) bands. The maximum at 6 eV binding energy clearly comes from the O(2*p*) state contribution while the smaller shoulder at 3 eV coincides with the N(2*p*) signal in UN sample.

In Ref. [23], the XPS and XRD methods as well as the measurement of ammonia concentration in the aqueous phase at the end of each experiment were used, in order to study corrosion of UN in water. $UO_2$ film arising during the surface reaction with water was detected using XPS for the surface of freshly polished UN pellet. The high corrosion rates of UN in water (at 928 °C) indicated that UN is not stable inside the hot aqueous environment. Corrosion rate for UN is much lower than that for U metal but higher that of uranium silicide.

Thickness, composition, concentration depth profile and ion irradiation effects on uranium nitride thin films deposited upon fused silica were investigated in [14] using Rutherford

Backscattering Spectroscopy (RBS) for 2 MeV $He^+$ ions. Deposition at -200 °C provided formation of thick stoichiometric UN film. This film was found to be stable for exposure to air. The surface oxidation is much more enhanced and the oxidized surface layer becomes gradually thicker in films deposited at higher temperature (+25 °C and +300 °C). A large influence of the ion irradiation on the film structure and layer composition was observed. This study also showed possibility to produce stoichiometric UN film with the required uranium content of 50% and to obtain the required film thickness by ion irradiation.

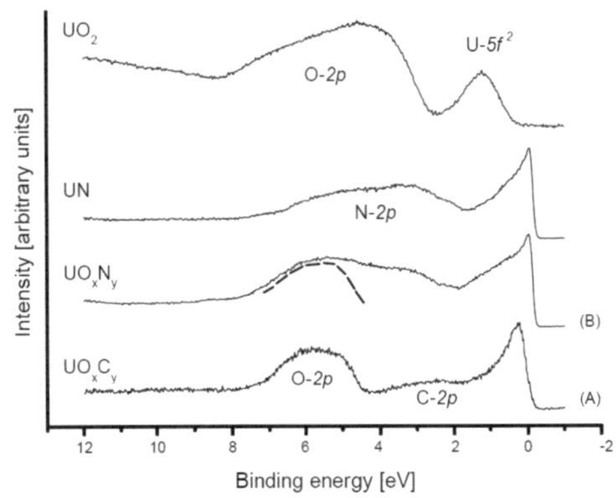


Fig. 2. He-II valence band spectra of $UO_xC_y$, $UO_xN_y$, UN and $UO_2$ spectra. Reproduced with permission from [22].

Finally, experimental studies also clearly showed that oxygen contacting to the surface of uranium mononitride can result in growth of the oxide compound and, at initial stages, can lead to the formation of surface layer structurally similar to oxynitrides $UO_xN_y$ [24].

### 2.3 Previous theoretical simulations on UN

Due to increasing interest in the fast breeder reactors and to the issues of transmutation of uranium, plutonium and minor actinides, first-principles and other theoretical calculations on actinide nitride compounds attract great attention nowadays. However, previous theoretical studies were performed mainly on UN bulk. Beginning from 80s [25-27], the methods based on the DFT were often applied to actinide materials.

In first relativistic calculations on UN single crystal, methods of full-potential Korringa-Kohn-Rostoker (KKR) Green's function [25] and Linear Muffin-Tin Orbitals (LMTO) [26-27] were used, focused mainly on the atomic and electronic structure. The calculated lattice parameters were found within 3% of experimental value, whereas the bulk modulus was reproduced worse when comparing with experimental data: by 23% higher [26] or within 10% [27]. The analysis of density of states (DOS) showed no gap between the valence and conduction bands in UN. The valence bands, found to be ~5-6 eV wide, appeared at ~2 eV below the Fermi level. The main peak was located by 1 eV below the Fermi level [27].

Recently, the all-electron calculations within the Linear Augmented Plane Wave (LAPW) approach were performed, using the PBE (Perdew-Burke-Ernzerhof) exchange-correlation functional (with and without incorporation of the spin-orbital coupling) as implemented in the WIEN-2k program package, for a series of actinide nitrides (AcN, ThN, PaN, UN, NpN,

PuN, AmN) [28]. The formation enthalpies mainly determined by the ground state cohesive energies were evaluated. The formation enthalpies are in excellent agreement with the experimental data (in the case of UN, the best correlation can be achieved with the results of calorimetric measurements: theoretical value of -291.0 $kJ\cdot mol^{-1}$ *vs.* experimental value of -290.5±1.4 $kJ\cdot mol^{-1}$ [29]). Some discrepancies between the experimental data and results of theoretical simulations for PuN and ThN still need to be clarified.

In Ref. [30], the same LAPW formalism within the GGA approximation was used to study the structural, electronic, and magnetic properties of the actinide compounds. The observed chemical bonding between the actinides and nitrogen was characterized by a significant ionic character. The calculated cohesive energies were found to be close to the experimental values (14.3 eV *vs.* 13.6 eV, respectively). Although the lattice constant for UN was calculated in a good agreement with the experiment (within ~0.4%), it was found to be ferromagnetic (FM) that contradicts to experimental results. The calculated spin density for UN in FM state was equal to 0.96 $\mu_B$. On the other hand, the calculated ferromagnetic structure of NpN and the non-magnetic structure of ThN agreed well with the corresponding experimental measurements.

In Ref. [31], the all-electron relativistic spin-polarized DFT calculations were performed, to evaluate the total energies, optimized geometries, as well as electronic and thermodynamic properties of perfect stoichiometric UN and $UN_2$ single crystals. For this purpose, the GGA Perdew-Wang (PW91) non-local exchange-correlation functional was used, and the numerical double-$\xi$ basis sets with *d*-type polarization functions were added to atoms heavier than hydrogen. Structural properties as measured using EXAFS and XRD methods were successfully reproduced in the calculations (within error of 0.03 Å). The DOS showed hybridization of the U(*6d*), U(*5f*) and N(*2p*) states as well as main contribution of *5f*-electrons to the conduction band. In this work the phonon frequencies and corresponding heat capacities were calculated. The authors suggested an important role of itinerant *5f* states in the thermodynamic properties.

The lattice parameters, electronic structure, as well as the thermodynamic properties of UN using LDA+*U* and GGA+*U* semi-empirical schemes and plane wave (PW) approach were presented in [32]. The total energy dependences on Hubbard *U*-parameter for UN bulk in FM and AFM states obtained in those calculations show that the FM state is preferable for the range of *U*-parameter between 0 and 2 eV while the AFM state could be favorable for *U*-parameter larger than 2 eV. Nevertheless, even though the AFM state of UN bulk is reproduced, the ground state is hardly obtainable when using the DFT+*U* method [33]. The value of *U*-parameter must be carefully suggested otherwise large errors may appear when calculating defect formation energies [34-35]. We avoid application of this method in the present study due to ferromagnetic nature of UN surface [36] reproducible by standart DFT functionals.

In other PW calculations on UN bulk, the *VASP* and *CASTEP* computer codes were employed using the PW91 exchange-correlation functional [37-38]. Both series of calculations agree well on the mixed metallic-covalent nature of UN chemical bonds reproducing the lattice constants, bulk moduli and cohesive energies.

The magnetic structure of UN was also addressed in Ref. [39]. In this study the so-called [111] magnetic structure was compared to the [001] one used here. Besides, the calculations were done within the carefull GGA+*U* study and did not reveal the energetic preference of the [111] magnetic structure except for very small values of Hubbard *U*-parameter.

Also, the PW approach combined with a supercell model was used for the calculations on defective UN crystal, containing single point defects as well as Frenkel and Schottky defect pairs. It was shown in Ref. [38] that the N-vacancies practically have no influence on the UN lattice constant, even for concentrations higher than 25%. The defect formation energies in the UN bulk were obtained to be equal 9.1-9.7 eV for the N-vacancy and 9.4-10.3 eV for the U-vacancy. The migration energy of the interstitial N-atom along the (001) axis was relatively low, i.e., 2.73 eV [37]. This fact confirms the suggestion that the interstitial migration might be a predominant mechanism of N-atom diffusion in the UN fuel [1]. Apart the behavior of empty vacancies, the O atom incorporation into vacancies in bulk UN was considered too [40]. Its incorporation into the N vacancies was found to be energetically more favorable as compared to the interstitial sites. However, the calculated values of solution energy showed an opposite effect. The calculated migration energy of the interstitial O atoms is very similar (2.84 eV). This fact confirms that the O atoms can easily substitute the host N atoms in UN structure.

## 3. Computational method

The UN (001) and (110) surfaces can be simulated using the symmetrical slabs containing odd number of atomic layers and separated by the vacuum gap of 38.9 Å corresponding, thus, to UN(001) 16 inter-layers. The vacuum gap is a property of plane wave approach. The suggested vacuum gaps are large enough to exclude the direct interaction between the neighboring two-dimensional (2D) slabs.

One should use the so-called supercell approach to simulate single point defects or an oxygen atom adsorbed/incorporated on/into the surface. Using 2x2 and 3x3 extenstions of primitive unit cell, lateral interactions between the defects can be also estimated. Such supercells contain four (2x2) and nine (3x3) pairs of the N- and U-atoms in each defectless layer of the slab. Periodically distributed surface vacancies (or oxygen atoms/molecules) *per* surface unit cell correspond to point defect (oxygen) concentrations of 0.25 and 0.11 monolayers (ML), respectively.

The results of first-principles PW calculations as obtained using Vienna Ab-intio Simulation Package (VASP) [41-43] will be further discussed. The VASP code treats core electrons using pseudopotentials, whereas the semi-core electrons of U atoms and all the valence electrons are represented by PWs. The electronic structure is calculated within the projector augmented wave (PAW) method [44]. Details on how the computation parameters could be properly chosen in such computations are discussed elsewhere [43]. Here we would like to mention that the cut-off energy in the calculations was chosen 520 eV. The integrations in the reciprocal space of the Brillouin zone were performed with 8x8x1 and 4x4x1 Monkhorst-Pack [45] mesh for the (001) and (110) surfaces, respectively. All the calculations involved the FM state of the surface only and full relaxation of all degrees of freedom if not otherwise stated.

## 4. Defect-free UN (001) and (110) surfaces

According to Tasker's analysis [46] the (001) surface must have the lowest surface energy for the rock-salt compounds. However, one could suppose facets with different crystallographic orientations for nano-particles and polycrystalline materials. Moreover, the role of different surfaces may be changed with the temperature. Therefore, additional calculations are also required for other surfaces to improve the validity of our results. In the present analysis we consider the (110) surface, for example. The (110) surface is characterized by smaller interlayer distances as compared to the (001) one. In this Chapter, the results of atomic oxygen adsorption, the formation of N-vacancies and oxygen atom incorporation, are discussed for the (110) surface and compared to those for the (001) surface.

The surface energy $E_{surf}$ as a function of the number of layers in the layers is given in Table 1 for both the (001) and (110) defectless surfaces (also shown in Fig. 3). The surface energy is calculated according to

$$E_{surf}(n) = \frac{1}{2S}(E_n - nE_b) \quad (1)$$

where $n$ the number of layers on the surface, $S$ the surface unit cell area, $E_n$ and $E_b$ the total energy of the surface unit cell and bulk primitive unit cell, respectively. The importance of spin relaxation is also addressed here. The spin magnetic moment relaxation leads to considerable changes (Table 1) suggesting lower $E_{surf}$ values and dependence on the number of layers in the slab. The lattice relaxation energies in spin-relaxed calculations turn out to be quite small, *i.e.*, ~0.03 eV. Depending on the slab thickness, $E_{surf}$ is ~0.5-0.7 J m$^{-2}$ larger for the (110) surface as compared to the (001) one. This further supports the importance of the (001) surface for this study.

The atomic displacements $\Delta z$ from perfect lattice sites differ significantly for the U atoms positioned at the surface and sub-surface layers (Table 2) being, however, somewhat larger for the 5-layer slab. The displacements of N atoms for all the slab thicknesses remain almost unchanged. Note that the N atoms on the (001) surface are displaced up whereas the U atoms are shifted inwards the slab center which results in the surface rumpling up to 1.2% of the lattice constant. In contrary, the surface U atoms of rumpled (110) surface lie higher than the corresponding N atoms.

In the next section we show how the electronic structure changes with the presence of point defects on the surface. This analysis will include the DOS and electronic charge distributions for perfect and defective surfaces.

## 5. Modeling of single N and U vacancies

### 5.1 Model and formation energies

To understand the oxidation mechanism in more detail, one has to take into account *surface defects* and their interaction with oxygen. The calculation of not only the surface defects, but also of the sub-surface and the central layer defects in the slab is necessary. Since we have chosen symmetrical slabs, two defects appear in the system due to symmetry with respect to the central layer. Let us define the formation energy of a point defect as

$$E_{form}^{\mathrm{N(U)}vac} = \frac{1}{2}\left(E^{\mathrm{UN(U/N\_vac)}} + 2E_{ref\_\mathrm{I(II)}}^{\mathrm{N(U)}} - E^{\mathrm{UN}}\right), \quad (2a)$$

for the surface and sub-surface vacancies, or

$$E_{\mathrm{form}}^{\mathrm{N(U)vac}} = E^{\mathrm{UN(U/N\_vac)}} + E_{ref\_\mathrm{I(II)}}^{\mathrm{N(U)}} - E^{\mathrm{UN}}, \tag{2b}$$

for a vacancy in the central layer of the slab. Here $E^{\mathrm{UN(U/N\_vac)}}$ the total energy of fully relaxed slab containing either the N or U-vacancies, $E^{\mathrm{UN}}$ the same for a defect-free slab, while $E_{ref\_\mathrm{I(II)}}^{\mathrm{N(U)}}$ is reference energy for the N (or U) atom.

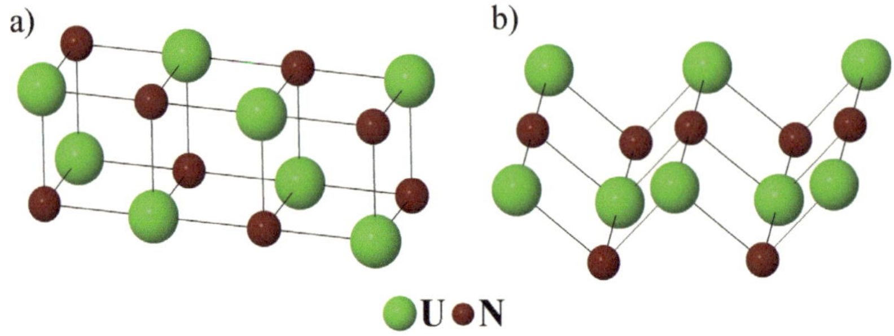


Fig. 3. The 2-layer model for the UN (001) (a) and (110) (b) surfaces.

| Number of layers | $E_{surf}$ (J m$^{-2}$) spin-frozen slab (001) | $E_{surf}$ (J m$^{-2}$) spin-relaxed slab (001) | $E_{surf}$ (J m$^{-2}$) spin-relaxed slab (110) |
|---|---|---|---|
| 5 | 1.69 | 1.44 | 1.977 |
| 7 | 1.70 | 1.37 | 1.928 |
| 9 | 1.70 | 1.29 | 1.878 |
| 11 | 1.69 | 1.22 | 1.830 |

Table 1. Surface energies $E_{surf}$ (J m$^{-2}$) for the defect-free UN (001) and (110) surfaces. In spin-frozen calculations, the spin magnetic moment of U atom was fixed at 1$\mu_B$

Two reference states for the calculations of the defect formation energy are often used in the literature which we would like to compare and discuss for our study. The first reference corresponds to the N(U) isolated atom in the triplet (quartet) spin state determined by $2p^3$ ($5f^36d^1$) valence electron configurations (hereafter, reference I in Table 3), *i.e.*,

$$E_{ref\_\mathrm{I}}^{\mathrm{N(U)}} = E_{atom}^{\mathrm{N(U)}}. \tag{3}$$

The isolated atom is calculated in a large rectangular parallelepiped box (28.28×28.28×22 Å$^3$).

The second reference state (hereafter, reference II as in Table 3) represents the chemical potential of N (U) atom which is defined as a function of temperature and partial nitrogen pressure. By neglecting these effects, the N chemical potential can be treated as the energy of atom in the molecule $N_2$ and thus considered at 0 K. Consequently, the chemical potential of U atom is given by the one-half total energy (*per* unit cell) of U single crystal in its low-temperature α-phase having the orthorhombic structure [47]. The corresponding second reference energies can be estimated as:

$$E_{ref\_II}^{\mathrm{N}} = \mu_{\mathrm{N}_2} = \frac{1}{2} E_{tot}[\mathrm{N}_2], \tag{3a}$$

$$E_{ref\_II}^{\mathrm{U}} = \mu_{\alpha\text{-U}} = \frac{1}{2} E_{tot}[\alpha\text{-U}], \tag{3b}$$

where $E_{tot}[N_2]$ is the total energy of nitrogen molecule while $E_{tot}[\alpha\text{-U}]$ the total energy of U metal bulk unit cell containing two atoms. In accordance with *Eqs. 3* the chemical potentials of N and U represent extreme cases of N(U)-rich conditions [48], *i.e.*, their minimum values have not been considered in the present study. The formation energy of N- (U-) vacancy with respect to the $N_2$ molecule (or α-U single crystal) and the energy of N (U) isolated atom are closely related: the former being larger than the latter by half the binding energy of the $N_2$ molecule or half the cohesive energy of α-U single crystal.

| Number of atomic planes | U | | | | N | | | |
|---|---|---|---|---|---|---|---|---|
| | (001) | | (110) | | (001) | | (110) | |
| | Surface | Sub-surface | Surface | Sub-surface | Surface | Sub-surface | Surface | Sub-surface |
| 5 | -0.050 | -0.012 | -0.053 | -0.005 | 0.023 | 0.023 | -0.279 | 0.068 |
| 7 | -0.046 | -0.009 | -0.038 | -0.009 | 0.024 | 0.028 | -0.272 | 0.092 |
| 9 | -0.047 | -0.011 | -0.042 | -0.014 | 0.024 | 0.028 | -0.279 | 0.091 |
| 11 | -0.047 | -0.011 | -0.015 | 0.015 | 0.025 | 0.031 | -0.252 | 0.118 |

* negative sign means an inward atomic displacement towards the mirror plane of the slab

Table 2. Atomic displacements $\Delta z$(Å)* for defect-free UN (001) and (110) surfaces.

The optimized lattice parameters of α-U ($a$ = 2.80 Å, $b$ = 5.88 Å, $c$ = 4.91 Å) have been slightly underestimated as compared to values obtained experimentally [47] and calculated [49-50], except for the parameter $b$ which is in a good agreement with experimental value of 5.87 Å [47] (while $a$=2.86 Å, $c$ = 4.96 Å [47]). Also, the ratios $c/a$, $b/a$ as well as the parameter $c$ are well verified by another plane-wave DFT study [51]. Analogously to an isolated nitrogen atom, the $N_2$ molecule has been calculated in the cubic box but of a smaller size (8×8×8 Å$^3$). The molecule $N_2$ is characterized by the bond length of 1.12 Å and the binding energy of 10.63 eV being well comparable with the experimental values of 1.10 Å and 9.80 eV [52], respectively. Note that the pre-factor of ½ in *Eq. 2a* arises due to a mirror arrangement of two N(U)-vacancies on the surface and sub-surface layers.

The formation energies of N- and U-vacancies ($E_{form}^{\mathrm{N(U)}\,vac}$), calculated for the two reference states as functions of the slab thickness, are collected in Table 3. These are smallest for the surface layer and considerably increase (by ~0.6 eV for the N-vacancy and by ~1.7 eV for the U- vacancy) in the sub-surface and central layers, independently of the reference state. This indicates the trend for vacancy segregation at the interfaces (surfaces or crystalline grain boundaries). A weak dependence of $E_{form}^{\mathrm{N(U)}\,vac}$ on the slab thickness is also observed. The value of $E_{form}^{\mathrm{N(U)}\,vac}$ is saturated with the slab thicknesses of seven atomic layers and more. Moreover, the difference between values of $E_{form}^{\mathrm{N(U)}\,vac}$ for the 5- and 7- layer slabs is less for the surface vacancies than for those in the central layer. This difference is the largest for the U-vacancy in the central plane (~0.16 eV).

The reference state II leads to smaller $E_{form}^{N(U)\,vac}$ (as compared to those found for the reference state I) and demonstrates a significant difference for two types of vacancies. According to reference II, the U vacancy could be substantially easier formed at T = 0 K than the N vacancy. Notice that the chemical potentials of O and U atoms used in similar defect studies in $UO_2$ bulk did not reveal the energetic preference for the U-vacancy [50, 53]. The defect-defect interaction is not responsible for this effect as $E_{form}^{N(U)\,vac}$ decreased by 0.1 eV only with the larger supercell size (3×3 in Table 3). On the other hand, the chemical potential of N may strongly depend on the temperature. For example, the formation energy of O-vacancy is reduced by almost 2 eV within a broad temperature range in perovskite oxides [54]. We, thus, expect that the trend will change of temperature effects are fully taken into account. Unlike the reference state II, the reference I results in similar formation energies for

| Layer | Number of atomic planes and supercell extension (in brackets) | Reference I | | Reference II | |
|---|---|---|---|---|---|
| | | U | N | U | N |
| Surface layer | 5 (2×2) | 8.63 | 8.84 | 1.46 | 3.70 |
| | 7(2×2) | 8.61 | 8.84 | 1.44 | 3.70 |
| | 9(2×2) | 8.61 | 8.84 | 1.44 | 3.71 |
| | 11(2×2) | 8.60 | 8.85 | 1.43 | 3.71 |
| | 5(3×3) | 8.51 | 8.78 | 1.34 | 3.64 |
| | 7(3×3) | 8.47 | 8.78 | 1.30 | 3.65 |
| Sub-surface layer | 5(2×2) | 10.31 | 9.38 | 3.14 | 4.25 |
| | 7(2×2) | 10.29 | 9.46 | 3.12 | 4.33 |
| | 9(2×2) | 10.26 | 9.46 | 3.09 | 4.33 |
| | 11(2×2) | 10.26 | 9.46 | 3.09 | 4.33 |
| | 7(3×3) | 10.18 | 9.47 | 3.01 | 4.34 |
| Central (mirror) layer[c] | 5(2×2) | 10.20 | 9.48 | 3.03 | 4.34 |
| | 7(2×2) | 10.36 | 9.57 | 3.19 | 4.43 |
| | 9(2×2) | 10.34 | 9.55 | 3.17 | 4.42 |
| | 11(2×2) | 10.39 | 9.56 | 3.22 | 4.42 |
| | 7(3×3) | 10.23 | 9.55 | 3.06 | 4.42 |

[a] reference energies I equal to -4.10 eV for U atom and -3.17 eV for N atom,
[b] reference energies II equal to -11.28 eV for U atom and -8.30 eV for N atom,
[c] defect formation energies for UN bulk using reference state I are 9.1-9.7 eV for the N- vacancy and 9.4-10.3 for the U-vacancy [38]

Table 3. The vacancy formation energies (in eV) for the two reference states (see the text for details).

both types of the vacancies. In the central slab layer, values of $E_{form}^{N(U)\,vac}$ are similar to those in the bulk (Table 3).

### 5.2 Surface reconstruction induced by vacancies

The local atomic displacements around the vacancies are largest for the nearest neighbors of vacancies. The analysis of atomic displacements allows us to suggest that the U-vacancy disturbs the structure of the surface stronger than the N-vacancy. If the N-vacancy lies in the surface layer, displacements of the nearest U atoms in the *z*-direction achieve 0.02-0.05 Å towards the central plane of the slab. The displacements of N atom nearest to the surface N vacancy achieve 0.05 Å towards the central plane (*z*-direction) and 0.01 Å in the surface plane (*xy*-displacement). Maximum displacements of neighbor atoms around the N-vacancy in the central plane have been found to be 0.04-0.07 Å (nearest U atoms from the neighboring layers are shifted in the *z*-direction towards the vacancy) not exceeding 0.025 Å for all the other atoms in the slab.

In contrast, the formation of U-vacancy results in much larger displacements of neighboring atoms, irrespectively of its position. If this vacancy lies in the surface layer, the displacements of 0.3-0.32 Å for the nearest N atoms are observed. If the U-vacancy lies in the central layer, the nearest N atoms from this layer are displaced by 0.17 Å while the N-atoms from the nearest layers are not shifted in *xy*-direction, being shifted by 0.15 Å towards the slab surface in the *z*-direction. The atomic displacements around the vacancies in the bulk have been found to be –0.03 Å and 0.13 Å for the N- and U-vacancies, respectively [38]. These values are close to those found in the calculations discussed here for the vacancies in the central slab layer, which mimics the bulk properties.

### 5.3 Electronic properties and finite-size effects

The finite slab-size effects caused by relatively large concentration of defects could be illustrated using the difference electron density redistribution $\Delta\rho(\mathbf{r})$. In Fig. 4, these redistributions are shown for N-vacancies positioned at both the outer (surface) and the central (mirror) planes of 5- and 7-layer slabs. Presence of two symmetrically positioned vacancies in the 5-layer slab induces their weak interaction across the slab (Fig. 4a) illustrated by appearance of an additional electron density around the N atoms in the central plane of the slab. Similarly, the vacancy in the mirror plane disturbs the atoms in the surface plane if thin slab contains only 5 layers (Fig. 4c). By increasing the slab thickness, we can avoid the effect of finite-slab size (Figs. 4b,d) which explains the stabilization of formation energies calculated for the 7-layer and thicker UN(001) slabs (Table 3).

The densities of states (DOS) are presented in Fig. 5. for both perfect and defective 7-layer UN slab. The U(5*f*) electrons are localized close to the Fermi level. These electrons are still strongly hybridized with the N(2*p*) electrons. It confirms the existence of covalent bonding observed in the analysis of Bader charges for defect-free surface. The N(2*p*) states form a band of the width ~4 eV, similar to that obtained in the bulk [5, 38]. In contrast, the contribution of U(6*d*) electrons remains insensitive to the presence of vacancies since the corresponding levels are almost homogeneously distributed over a wide energy range including the conduction band.

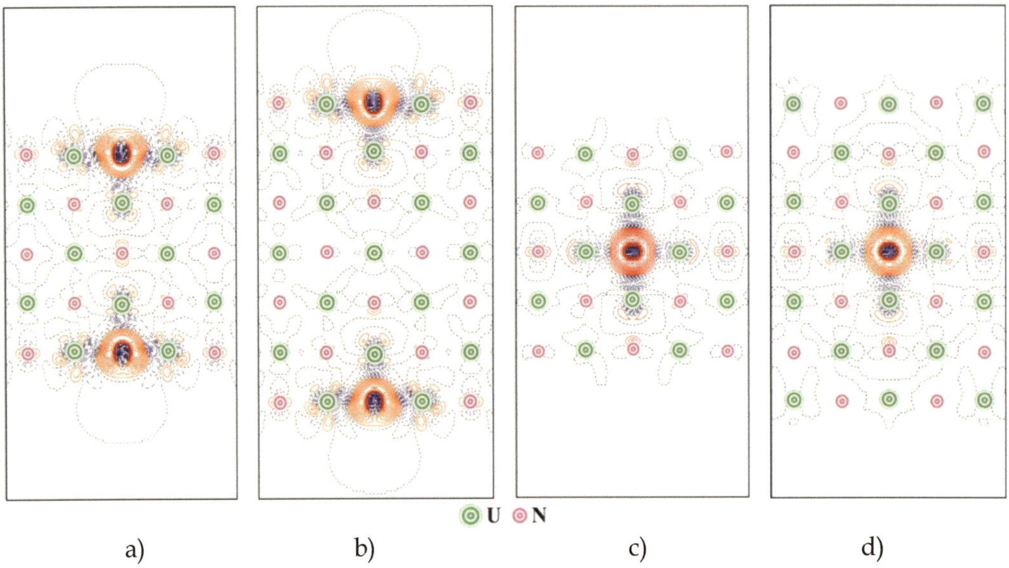


Fig. 4. The 2D sections of the electron density redistributions around the N-vacancies in the 5- and 7-layer (001) slabs (2×2) defined as the total electron density of defective surface minus a superposition of the electron densities for both perfect surface and isolated atom in the regular position on the surface: a) the N-vacancy in the surface plane, the 5-layer slab, b) the same for the 7-layer slab, c) the N-vacancy in the central plane, the five-layer slab, d) the same for the 7-layer slab. Solid (red) and dashed (blue) isolines correspond to positive (excess) and negative (deficiency) electron density, respectively. Isodensity increment is 0.0025 $e \cdot$a.u.$^{-3}$.

### 5.4 Comparison of the UN (001) and (110) surfaces

To increase the reliability of the results we compare also the results of point defect calculations in the surface layer of the (001) and the (110) surfaces (Table 4). Let us consider the 5-, 7-, 9-, and 11-layer 2×2 surface supercells as well as 7-layer 3×3 supercell for the (110) surface. The N-vacancy formation energies are by ~0.7 eV smaller for the (110) surface. One can explain it due to a larger friability of the (110) surface as compared to the (001) surface. The dependence on the slab thickness is more pronounced for the (001) surface: the formation energy of N-vacancy increases with the thickness by 0.012 eV.

| Number of layers and supercell size | (001) | (110) |
|---|---|---|
| 5, 2×2 | 3.700 | 3.075 |
| 7, 2×2 | 3.706 | 3.028 |
| 9, 2×2 | 3.708 | 3.036 |
| 11, 2×2 | 3.712 | 3.026 |
| 7, 3×3 | 3.646 | 2.966 |

Table 4. The N-vacancy formation energies (in eV) evaluated for the (001) and (110) surfaces.

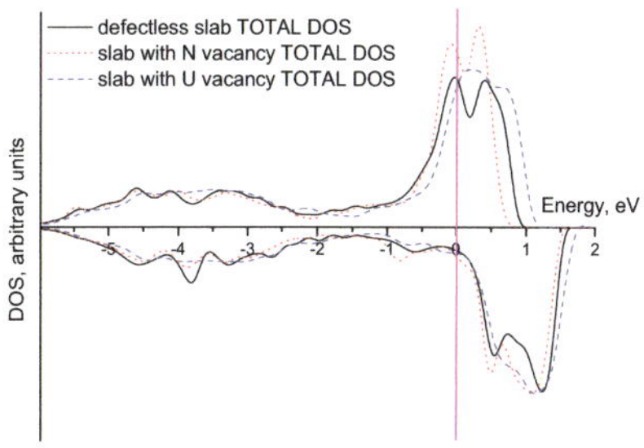


a)

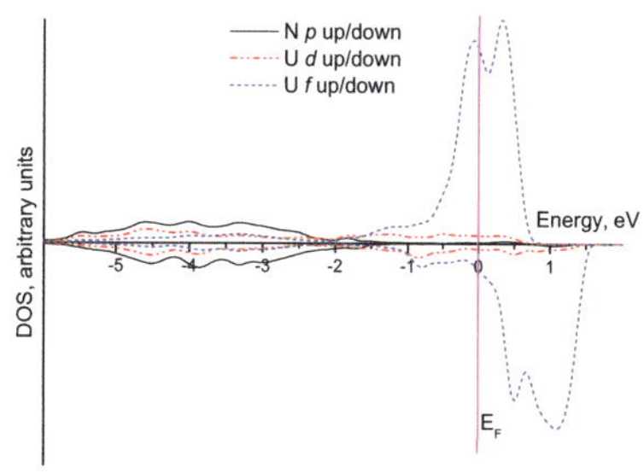


b)

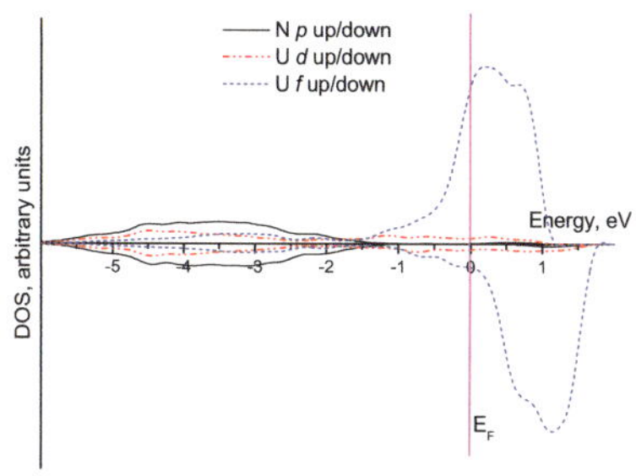


c)

Fig. 5. The total and projected DOSs of the 7-layer (001) surface (2×2):
a) total DOS of defective and defect-free surfaces
b) projected DOSs for the surface containing the N-vacancies
c) projected DOSs for the surface containing the U-vacancies.

## 6. Molecular oxygen adsorption

### 6.1 Model and binding energy

The interaction of *molecular* oxygen $O_2$ with the perfect UN(001) surface [7] and its dissociation on metallic UN surface will be considered here. It is impotant to consider prior simulations of O atom adsorption. An important issue for these interactions would be whether the $O_2$ dissociation upon the surface is energetically possible, which adsorption sites are optimal for this, and whether it can occur spontaneously, without energy barrier, similarly to other metallic surfaces, for example Al [55].

Results as discussed here for molecular adsorption were performed using the fixed spin magnetic moment of U at 1 $\mu_B$. We rely on this approximation, which allowed us to speed the calculations substantially up. In these simulations the 5-layer slab with the 2×2 supercell was used only. The periodic adsorbate distribution corresponds to the molecular coverage of 0.25 ML (or atomic O coverage of 0.5 ML). The binding energy $E_{bind}$ *per* oxygen atom in the adsorbed molecule $O_2$ was calculated as:

$$E_{bind} = \frac{1}{4}\left(E^{\mathrm{UN}} + 2E^{\mathrm{O_2}} - E^{\mathrm{O_2/UN}}\right), \tag{4}$$

where $E^{O_2/UN}$ the total energy of a fully relaxed $O_2$/UN(001) slab for several configurations of ($O_2$) upon the surface (Fig. 6), $E^{O_2}$ and $E^{UN}$ the total energies of an isolated oxygen molecule in the ground (triplet) state and of a relaxed defectless UN slab, respectively. The pre-factor 1/4 appears due to the symmetrical slabs containing two equivalent surfaces with adsorbed $O_2$. Note that each molecule before and after dissociation contains two O atoms. When modeling the molecular adsorption, different configurations of the $O_2$ molecule in the triplet state on surface are possible. *Vertical* orientations of the molecule were found to be unstable. But, the *horizontal* configurations suggest stable configurations. The binding energy of the molecule, using Eq. (4), and its dissociation energy (representing the difference of the total energies of a slab with an $O_2$ molecule before and after dissociation) are given in Table 5.

### 6.2 Spontaneous dissociation

A spontaneous barrierless $O_2$ dissociation indeed takes place in the two following cases, when the molecular center is atop either (*i*) a hollow site or (*ii*) N atom, with the molecular bond directed towards the two nearest U atoms (the configurations 1 and 5 in Fig. 6, respectively). The relevant dissociation energies $E_{diss}$ are given in Table 5 together with other parameters characterizing the atomic relaxation and the Bader charge distribution. Geometry and charges for these configurations after dissociation are close to those obtained for chemisorbed O atoms (see the next section, too), i.e., the surface U atoms beneath the oxygen adatom after dissociation are shifted up in both configurations (Table 5). However, since concentration of O is twice as larger as compared to that for the atomic adsorbtion [6,10], some differences of the results still occur. These may be characterized by the *repulsion* energy of ~0.7 eV between the two adatoms after $O_2$ dissociation, which are positioned atop the two nearest U atoms (the configuration 1). Two more configurations of adsorbed $O_2$ are possible, i.e. the dissociation is energetically possible with energy barrier: (*i*) atop the hollow site when a molecular bond is oriented towards the nearest N atoms (the configuration 2 in Fig. 6) and (*ii*) atop the U atom (for any molecular orientation, *e.g.*, the configurations 3 and

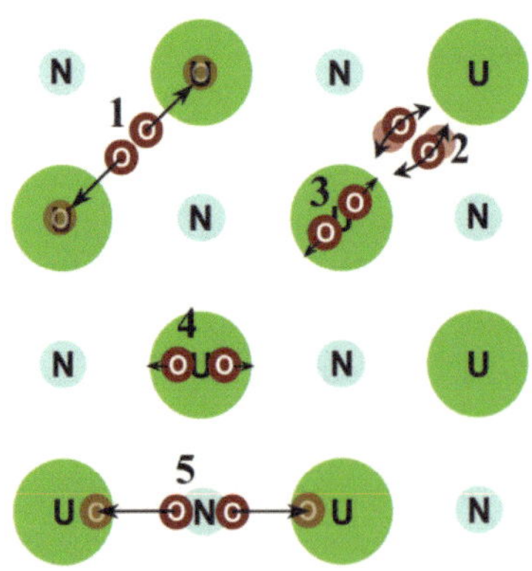


Fig. 6. Schematic view of five different horizontal configurations for the $O_2$ molecule adsorption on UN (001) surface: 1) atop the hollow site between U atoms 2) atop the hollow site between N atoms 3) atop U atoms oriented towards the next-nearest surface U atoms 4) atop the U atoms oriented towards the nearest N atoms 5) atop the N atoms oriented towards the nearest U atoms. We show that spontaneous dissociation of molecule can occur when $O_2$ is located either atop the hollow site (1) or atop the N atom (5).

4 in Fig. 6). For the configuration 2, the molecule rotates to take a stable position between the U atoms with further dissociation. The configurations 3 and 4 rather describe metastable $UO_2$ *quasi-molecules*, due to a strong bonding between all three atoms, and, also because the corresponding U atom is noticeably shifted up from its initial positions on the surface. The dissociation of the $O_2$ molecule in configuration 3 is energetically possible but only after overcoming the activation energy barrier.

| Configuration | | $E_{bind}$ | $z^a$ | $E_{diss}$ | $\Delta z^e$(U1) | $\Delta z^e$(U2) | $\Delta z^e$(N) |
|---|---|---|---|---|---|---|---|
| (1) | before dissociation | 3.03 | 1.893 | - | -0.050 | -0.050 | 0.025 |
| | after dissociation | 6.04 | 1.957 | 3.01 | 0.075 | 0.068 | -0.133 |
| atop U | (3) | 4.00 | 2.18 | - | 0.176 | -0.048 | -0.096 |
| | (4) | 4.18 | 2.14 | - | 0.123 | -0.051 | -0.106 |
| (5) | before dissociation | 2.67 | 2.020 | - | -0.050 | -0.050 | 0.025 |
| | after dissociation | 5.85 | 1.955 | 3.18 | 0.073 | 0.021 | -0.201 |

[a] $z$ is the height of O atoms with respect to the non-relaxed UN surface
[e] $\Delta z$ the additional vertical shifts of the same surface atoms from their positions in absence of adsorbed oxygen.

Table 5. The calculated values of binding ($E_{bind}$ *per O atom in eV*) and dissociation ($E_{diss}$ *in eV*) energies as well as geometry (*z, Δz in Å*) for molecular configurations and those for spontaneous dissociative chemisorption of $O_2$. The configurations are shown in Fig. 6. The calculated binding energy for a free $O_2$ molecule in the triplet state is 6.06 eV and a bond length is 1.31 Å (*cf.* with experimental values of 5.12 eV and 1.21 Å, respectively) [56].

## 7. Modeling of O atom adsorption and migration on perfect UN surface

### 7.1 Model and binding energies

To simulate the O atom adsorption, we consider the 5- and 7-layer slabs for the (001) surface with the 2×2 and 3x3 supercell sizes (Figs 7-8). The binding energy $E_{bind}$ of adsorbed oxygen atom is calculated with respect to a free O atom:

$$E_{bind} = \frac{1}{2}\left(E^{\mathrm{UN}} + 2E^{\mathrm{O_{triplet}}} - E^{\mathrm{O/UN}}\right), \qquad (5)$$

where $E^{\mathrm{O/UN}}$ the total energy of relaxed O/UN(001) slab for $O_{ads}$ positions atop either the N or U surface ions, $E^{\mathrm{O_{triplet}}}$ and $E^{\mathrm{UN}}$ the energies of an isolated O atom in the ground (triplet) state and of a relaxed defectless UN slab. The free O atom is calculated in the cubic box with the edge of ~20 Å. The pre-factor ½ appears since the surface is modeled by a slab with two equivalent surfaces and O is positioned symmetrically with respect to the central layer in the slab. We also can estimate $E_{bind}$ in the case of the defective surface with one N vacancy in the surface layer according to

$$E_{bind} = \frac{1}{2}\left(E^{\mathrm{UN(N\_vac)}} + 2E^{\mathrm{O_{triplet}}} - E^{\mathrm{O/UN(N\_vac)}}\right), \qquad (6)$$

where $E^{\mathrm{UN(N\_vac)}}$ the total energy of defective UN substrate containing the N vacancy while $E^{\mathrm{O/UN(N\_vac)}}$ the total energy of adsorbed oxygen atoms atop the defective substrate (Table 6).

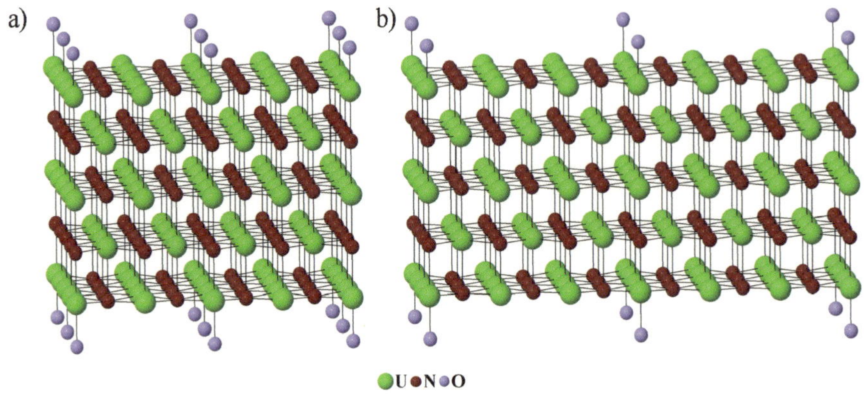


Fig. 7. Model of O/UN(001) interface: two-sided adsorption of O atoms regularly distributed atop $U_{surf}$ atoms with the 2×2 (a) and 3×3 (b) periodicity

Due to a mixed metallic-covalent nature of the chemical bonding in UN, we expect a high affinity of adsorbed O towards the UN(001) surface. The binding energy *per* O adatom is expected to be similar to that on regular O/Al(111) and/or O/Al(001) metallic interfaces (~10 eV) [55] unlike that on semiconducting O/$SrTiO_3$(001) interfaces (~2 eV) [57]. Indeed, the $E_{bind}$ values of 6.9-7.6 and 5.0-5.7 eV *per* O adatom atop the surface U or N atoms, respectively, are accompanied by 0.5-1.2 *e* charge transfer from the surface towards the O adatom (Tables 6 and 7). The positively charged surface U atom goes outwards, minimizing its distance with the adsorbed O atom (Fig. 9). The N atom is strongly displaced from the adsorbed O atom inwards the slab, due to a mutual repulsion between N and O. Tables 6 and 7 clearly demonstrate the ionic character of surface O–U bond.

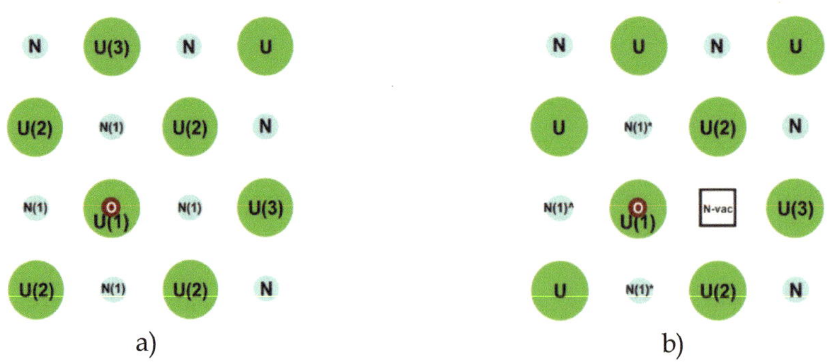


Fig. 8. Schematic top view of O adatoms located atop the surface U atom without (a) and with (b) N vacancy in the proximity of adsorbed O. Numbers in brackets enumerate non-equivalent surface atoms described in Tables 6 and 7.

Electron density redistributions caused by the absorption of O atom atop N or U atoms of the (001) surface are shown in Fig. 9. An analysis of the difference density plots for both configurations of O confirms that the oxygen adatom forms a strong bonding with the U atom which can be considered as one-center adsorption complex (Fig. 9c, 9d). In the case of O adatom atop the N atom, O atoms forms a multi-center adsorption complex involving four adjacent surface U atoms (Fig. 9a, 9b). As follows from Table 7, these surface atoms mostly contribute to the high O binding energy atop N. Formation of the strong chemical bonding of O atom with U results in a strong anisotropic redistribution of the electronic charges, thus, indicating considerable contribution of U 5*f*- and 6*d*-electrons to chemical bonding.

| Model | $E_{bind}$ | $q_O$ | $q_{U(1)}$ | $q_{U(2)}$ | $q_{U(3)}$ | $q_{N(1)}$ | $d_{O-U}$ | $\Delta z_{U(1)}$ | $\Delta z_{U(2)}$ | $\Delta z_{U(3)}$ | $\Delta z_{N(1)}$ |
|---|---|---|---|---|---|---|---|---|---|---|---|
| 2×2 5-layers | 7.57 | -1.08 | 2.09 | 1.82 | 1.84 | -1.63 | 1.88 | +0.16 | +0.025 | +0.003 | -0.09 |
| 2×2 7-layers | 7.51 | -1.08 | 2.19 | 1.78 | 1.78 | -1.64 | 1.89 | +0.17 | +0.03 | -0.02 | -0.09 |
| 2×2 7-layers & N-vacancy | 7.58[b] | -1.08 | 1.84 | 1.50 | 1.48 | -1.61*<br>-1.61^ | 1.88 | +0.14 | +0.01 | -0.02 | -0.09*<br>-0.08^ |
| 3×3 5-layers | 7.59 | -1.09 | 2.13 | 1.80 | 1.74 | -1.62 | 1.88 | +0.16 | +0.01 | -0.01 | -0.10 |
| 3×3 7-layers | 7.57 | -1.09 | 2.13 | 1.78 | 1.79 | -1.62 | 1.88 | +0.16 | +0.01 | -0.01 | -0.09 |
| 3×3 7-layers & N-vacancy | 7.59[b] | -1.09 | 1.86 | 1.47* | 1.38* | -1.61*<br>-1.61^ | 1.88 | +0.10 | -0.025* | -0.06* | -0.12*<br>-0.11^ |

[a] positive sign corresponds to atom displacement outward the substrate; *,^ adsorbed O atom in the presence of non-equivalent N atoms for system with the N-vacancy

Table 6. The binding energy ($E_{bind}$ *in eV*), the equilibrium distance between O and surface U atom ($d_{O-U}$ in Å), the effective atomic charges of atoms ($q$ *in* $e^-$), and vertical U and N atoms displacements ($\Delta z$)[a] from the surface plane for adatom position atop U. Values of $q$ for the surface atoms on the perfect surface equal +1.68 (+1.74) $e$ for U and –1.65 (–1.67) $e$ for N [8].

| Model | $E_{bind}$ | $q_O$ | $q_{N(1)}$ | $q_{N(2)}$ | $q_{N(3)}$ | $q_{U(1)}$ | $d_{O-N}$ | $\Delta z_{N(1)}$ | $\Delta z_{N(2)}$ | $\Delta z_{N(3)}$ | $\Delta z_{U(1)}$ |
|---|---|---|---|---|---|---|---|---|---|---|---|
| 2×2 5-layers | 5.52 | -1.17 | -1.48 | -1.68 | -1.68 | 1.86 | 2.19 | -0.69 | +0.03 | +0.05 | +0.13 |
| 2×2 7-layers | 5.58 | -1.17 | -1.48 | -1.63 | -1.67 | 1.86 | 2.21 | -0.715 | +0.03 | +0.03 | +0.12 |
| 3×3 5-layers | 5.57 | -1.18 | -1.51 | -1.67 | -1.68 | 1.89 | 2.20 | -0.70 | +0.01 | +0.01 | +0.13 |
| 3×3 7-layers | 5.65 | -1.18 | -1.51 | -1.69 | -1.65 | 1.89 | 2.22 | -0.73 | +0.01 | +0.02 | +0.12 |

[a] atomic positions of U and N ions are reversed as compared to those shown in Fig. 8a.

Table 7. The calculated parameters for O atom adsorbed atop N[a] (see caption and footnotes of Table 6 for explanation).

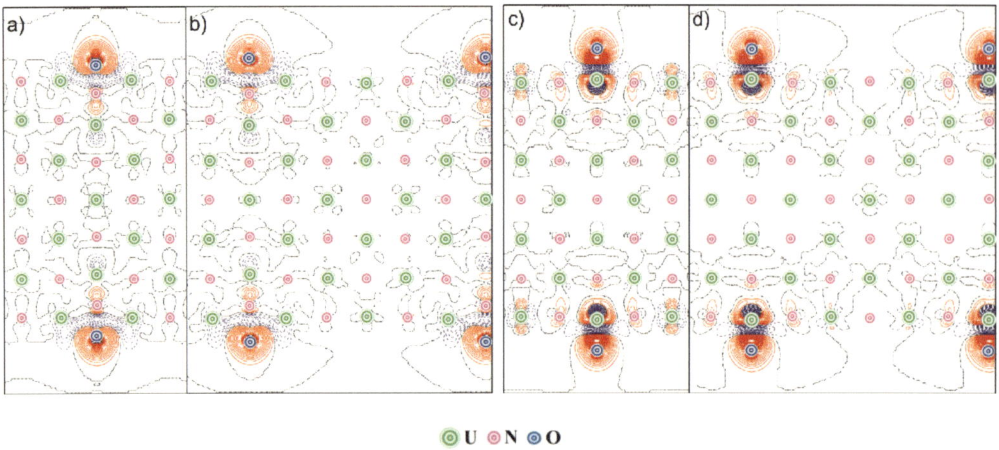


Fig. 9. The 2D sections of the electron charge density re-distributions $\Delta\rho(\mathbf{r})$ for O atoms adsorbed atop (*i*) N atom for 2×2 (a) and 3×3 (b) supercells as well as (*ii*) U atom for 2×2 (c) and 3×3 (d) supercells upon the seven-layer UN(001) slab. Function $\Delta\rho(\mathbf{r})$ is defined as the total electron density of the interface containing adsorbed O atom minus the densities of substrate and adsorbate with optimized interfacial geometry. Solid (red) and dashed (blue) isolines correspond to positive and negative electron densities, respectively. Dot-dashed black isolines correspond to the zero-level.

Plots of electronic density redistributions clearly show that the U atoms shield influence of neighbor atoms on the next coordination spheres much better than the N atoms.

From the viewpoint of finite size effects, the 3×3 supercell allows one to reduce the interactions between the adsorbed O atoms (as well vacancies if present). As seen in Table 7. the charges are slightly larger and displacements are smaller for larger supercells. The choice of supercell size influences the interactions between the defects across the slab. In particular, adsorption atop the N atom is sensitive to this effect.

### 7.2 Comparison of the UN (001) and (110) surfaces

The binding energies of oxygen adatom with UN(110) surface are given in Table 8. For both the surfaces, the binding energies with the U atom are larger as compared with the N atom (~1.9 eV for the (001) vs. ~2.1-2.2 eV for the (110) surface). Moreover, if the supercell size is

| Number of layers and supercell size | | U | N |
|---|---|---|---|
| (001) | 7, 2×2 | 7.51 | 5.58 |
| | 7, 3×3 | 7.57 | 5.65 |
| (110) | 7, 2×2 | 7.90 | 5.73 |
| | 7, 3×3 | 7.91 | 5.99 |

Table 8. The calculated binding energies ($E_{bind}$, eV) for oxygen atom adsorption atop UN (001) and (110) surfaces.

increased from 2×2 to 3×3, then $E_{bind}$ is also increased. The $E_{bind}$ values on the (110) surface are ~0.1-0.4 eV larger as compared to the (001) one. Such higher $E_{bind}$ values for the (110) surface could be explained by larger distances between the surface adatoms upon the (110) surface resulting in decreased lateral interactions between the adsorbed O atoms.

### 7.3 Migration path for O ad-atom

Three main migration paths of O upon the UN (001) surface (Fig. 10) are as follows [10]: (*i*) path 1: between U atom and the nearest N atom, (*ii*) path 2: between the two neighboring U atoms, (*iii*) path 3: between neighboring N atoms.

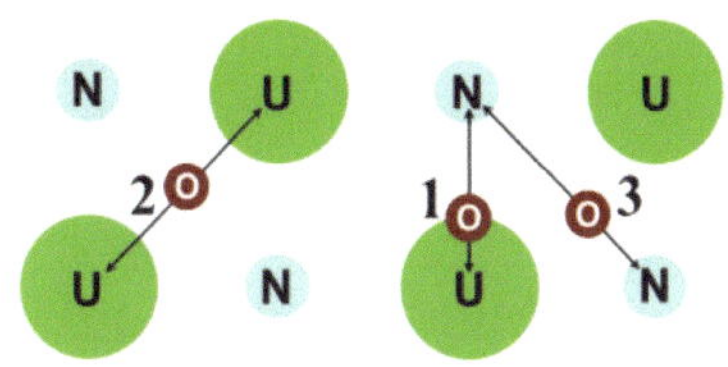


Fig. 10. Different oxygen migration paths upon the UN(001) surface (top view).

| 1. From site atop $U_{surf}$ to site atop N (migration path 1) | | | | |
|---|---|---|---|---|
| *Supercell size:* | 2×2 | | 3×3 | |
| *Number of atomic layers:* | 5 | 7 | 5 | 7 |
| atop U | 7.57 | 7.51 | 7.59 | 7.57 |
| ¼ of distance U-N (or 0.61 Å from U atom) | 7.39 | 7.39 | - | - |
| ½ of distance U-N (or 1.22 Å from U atom) | 6.97 | 6.98 | - | - |
| ¾ of distance U-N (or 1.83 Å from U atom) | 5.91 | 5.93 | - | - |
| atop $N_{surf}$ (or 2.43 Å from U atom) | 5.52[b] | 5.58 | 5.57 | 5.65 |
| 2. From hollow position (h.p.) to site atop U (migration path 2) | | | | |
| *Supercell size:* | 2×2 | | 3×3 | |
| *Number of atomic layers:* | 5 | 7 | 5 | 7 |
| atop h.p. | 7.21 | 7.245 | 7.20 | 7.21 |
| ¼ of distance h.p.-U (or 0.43 Å from h.p.) | 7.23 | 7.255 | - | - |
| ½ of distance h.p.-U (or 0.86 Å from h.p.) | 7.32 | 7.33 | - | - |
| ¾ of distance h.p.-U (or 1.29 Å from h.p.) | 7.45 | 7.45 | - | - |
| atop $U_{surf}$ (or 1.72 Å from h.p.) | 7.57 | 7.51 | 7.59 | 7.57 |
| 3. From hollow position (h.p.) to site atop $N_{surf}$ (migration path 3) | | | | |
| *Supercell size:* | 2×2 | | 3×3 | |
| *Number of atomic layers:* | 5 | 7 | 5 | 7 |
| atop h.p. | 7.21 | 7.25 | 7.20 | 7.21 |
| 1/4 of distance h.p.-N (or 0.43 Å from h.p.) | 6.61 | 6.65 | - | - |
| 1/2 of distance h.p.-N(or 0.86 Å from h.p.) | 6.32 | 6.35 | - | - |
| 3/4 of distance h.p.-N(or 1.29 Å from h.p.) | 5.54 | 5.57 | - | - |
| atop $N_{surf}$ (or 1.72 Å from h.p.) | 5.52 | 5.58 | 5.57 | 5.65 |

Table 9. Binding energies $E_{bind}$ of O atoms in different positions atop UN slab (Fig. 10).

In contrast to adsorption calculations, only z coordinates of all atoms in the slab were relaxed in these migration calculations. The results of O migration for different slab thicknesses and supercell extensions are summarized in Table 9. These represent the values of binding energies calculated for migration paths of O adatoms upon the perfect UN(001) substrate shown in Fig. 10. We fix five sites along the O migration trajectories for 2×2 supercells of UN(001) slab and two sites for 3×3 supercells. In both cases, the most favorable migration trajectory has been optimized to be the line joining the sites atop the nearest surface U atoms and the hollow sites between them (trajectory 2). The corresponding energy barriers found (0.36 eV for the 5-layer slab and 0.26 eV for the 7-layer slab) indicate a high mobility of adsorbed O atoms upon UN. The energy barriers along other two migration trajectories are substantially larger (1.93-2.05 eV and 1.31-1.69 eV for trajectories 1 and 3 shown in Fig. 10). Thus, we predict quite high mobility of atoms along the surface, due to relatively low migration barriers.

## 8. Oxygen migration and incorporation into the surface vacancies

### 8.1 Model of ad-atom migration

To estimate the oxygen adatom mobility upon the defective UN(001) surface, we also have performed a series of calculations of O atom adsorbed atop the surface U atom in the proximity of the surface N-vacancy (Fig. 8b). According to our calculations, this O atom can be captured by the vacancy when overcoming a low energy barrier (~0.5-1 eV). We have estimated the energy gain for such a transition of oxygen adatom using the formula:

$$\Delta E_g = \frac{1}{2}\left(E_{tot}^{\mathrm{UN(O\_in\_N\_vac)}} - E_{tot}^{\mathrm{UN(O\_atop\_U)}}\right), \tag{7}$$

where $E_{tot}^{\mathrm{UN(O\_in\_N\_vac)}}$ is the total energy of the supercell containing the O atom in the N-vacancy, and $E_{tot}^{\mathrm{UN(O\_atop\_U)}}$ the total energy of the supercell with O atom adsorbed atop U atom positioned in the proximity of existing N-vacancy. For calculations on the total energies in Eq. 7 we have fixed horizontal $x$ and $y$ oxygen coordinates, to prevent the O adatom migration. The pre-factor ½ in Eq. (7) appears due to the symmetric arrangement of adsorbed or incorporated O atoms. The calculated energy gain ($\Delta E_g$) for the transition from position atop U atom to position in the N-vacancy equals to ~2 eV *per* oxygen adatom (1.99 eV for 2×2 7-layer supercell and 1.94 eV for 3×3 7-layer supersell). Thus, we have showed the possibility of low-barrier oxygen adatom incorporation into existing the N-vacancy from the nearest adsorption site atop the U atom.

### 8.2 Oxygen incorporation and solution energies

The incorporation of O atom into the surface vacancies is expected along with oxygen atom diffusion along the surface or right after $O_2$ dissociation. One of possible ways for UN surface oxidation is the formation of oxynitride islands or films upon the UN surface [22]. The energies (of incorporation $E_I$ and of solution $E_S$ into the (001) surface) which characterize this process are discussed here. Hence, it is very impotant to describe the oxygen interaction with the single vacancies. As known from the literature, considerable attention was paid so far to the static and dynamic properties of primary defects (vacancies and incorporated impurities) in UN *bulk* [38]. These defects affect the fuel performance during operation and its reprocessing. Apart from the behavior of empty vacancies, the O atom incorporation into them in bulk UN was also considered. Incorporation of O into the N-vacancy in the bulk was found to be energetically more favorable in comparison with the interstitial sites [58]. However, $E_S$ demonstrates an opposite behavior.

Our calculations have been performed for surface 2×2 and 3×3 UN supercells. The O atom can occupy either the N- or U-vacancies in the surface, sub-surface and central layers of the slab. Due to the presence of mirror layers in the symmetric slabs, one can consider the two-sided symmetric arrangement of defects.

The energy balance for the incorporation of an O atom into a vacancy can be characterized by *the incorporation energy* $E_I$ as suggested by Grimes and Catlow [59]

$$E_I = E^{\mathrm{UN(O\_}inc)} - E^{\mathrm{UN(N/U\_vac)}} - E^{O}, \quad (8a)$$

for the O atom incorporated into the N- and U vacancy in the central atomic layer and

$$E_I = \frac{1}{2}(E^{\mathrm{UN(O\_}inc)} - E^{\mathrm{UN(N/U\_vac)}} - 2E^{O}), \quad (8b)$$

for the same incorporation in the surface or sub-surface layers. Here $E^{\mathrm{UN(O\_}inc)}$ the total energy of the supercell containing the O atom at either the N- or U-vacancy ($E^{\mathrm{UN(O\_}inc)} < 0$), $E^{\mathrm{UN(N/U\_vac)}}$ the energy of the supercell containing an unoccupied (empty) vacancy, and $E^{O}$ half the total energy of isolated $O_2$ molecule in the triplet state (representing the oxygen chemical potential of O at 0 K). Since the value of $E_I$ describes the energy balance for the incorporation into pre-existing vacancies, it has to be negative for energetically favorable incorporation processes.

To take into account the total energy balance, including the vacancy formation energy $E_{form}$ in the defect-free slab, the solution energy [59] is defined as:

$$E_S = E_I + E_{form}, \quad (9)$$

where $E_{form}$ the formation energy of N- or U-vacancy in the slab calculated using *Eqs. 2a* and *2b*. The parameters and properties of calculated $O_2$ molecule and α-U are the same as discussed in previous sections.

The calculated O adatom incorporation into the N-vacancy of the UN(001) surface has been found to be energetically favorable since both values of $E_I$ and $E_S$ are strictly negative (Table 10). This is in favor of both creation of the N vacancy and adsorption of the O atom from air. Also, $E_I$ decreases by ~0.4 eV (becomes more negative) within the slab as compared to the surface layer, whereas $E_S$ is smallest for the N-vacancy just on the surface layer. Contrary, the values of $E_I$ for the surface and central layers have been found to be close to zero in case of U-vacancy. The sub-surface layer is characterized by ~1 eV smaller values of $E_I$ than for the surface and central layers. Our results indicate importance of oxynitride formation. However, $E_S$ is positive and increases for O atoms in the U-vacancy and the slab center.

Table 10 also indicates that solution of oxygen atoms is energetically more favorable at the surface layers than inside the slab. As the supercell size increases (see the 3×3 extension in Table 10), both $E_I$ and $E_S$ decrease whereas influence of the slab thickness is not so clear. Nevertheless, the U-vacancy appeared to be most sensitive to the supercell size related to spurious interactions between the periodically repeated defects. The $E_I$ as well as $E_S$ values may be reduced by 0.15 eV at the average in this case.

| Layer | Supercell size | Number of layers | N | | | U | | |
|---|---|---|---|---|---|---|---|---|
| | | | $E_I$ | $E_S$ | $q_{eff}$ | $E_I$ | $E_S$ | $q_{eff}$ |
| Surface | 2×2 | 5 | -6.17 | -2.47 | -1.36 | -0.34 | 1.12 | -0.98 |
| | | 7 | -6.18 | -2.48 | -1.36 | -0.86 | 0.58 | -1.03 |
| | | 9 | -6.19 | -2.48 | -1.36 | -0.94 | 0.49 | -1.06 |
| | 3×3 | 5 | -6.12 | -2.48 | -1.37 | -0.68 | 0.654 | -1.05 |
| | | 7 | -6.13 | -2.48 | -1.36 | -1.07 | 0.230 | -1.08 |
| Subsurface | 2×2 | 5 | -6.31 | -2.07 | -1.42 | -1.86 | 1.284 | -1.10 |
| | | 7 | -6.42 | -2.09 | -1.40 | -1.82 | 1.297 | -1.10 |
| | | 9 | -6.42 | -2.09 | -1.40 | -1.82 | 1.271 | -1.10 |
| | 3×3 | 7 | -6.43 | -2.09 | -1.39 | -2.01 | 1.000 | -1.10 |
| Central (mirror) | 2×2 | 7 | -6.61 | -2.18 | -1.42 | 0.74 | 3.923 | -0.89 |
| | | 9 | -6.61 | -2.19 | -1.38 | 0.67 | 3.838 | -0.90 |
| | 3×3 | 7 | -6.60 | -2.18 | -1.42 | 0.32 | 3.378 | -0.94 |

Table 10. The incorporation ($E_I$) and solution ($E_S$) energies (in eV), effective charge of oxygen atoms ($q$ in $e^-$) for the O incorporation. The reference states for incorporation and solution energies into the U- and N-vacancies are the chemical potentials of O, N and U (see the text for details).

### 8.3 Electronic properties and finite-size effects

Large concentrations of defects (25% for the 2×2 extension in Table 10) causes certain finite-size effects which can be illustrated using the 2D difference electron density redistributions $\Delta\rho(\mathbf{r})$. These plots are shown for the O atoms incorporated into the N-vacancies at the surface (Fig. 11). Inside the 5-layer slab, a presence of the two symmetrically positioned defects induces their interaction (visible in charge redistribution across a slab in Fig. 11a). An increase of the slab thickness reduces this effect (Fig. 11c). If the supercell size is decreased (the 2×2 supercell, Fig. 11b) an additional electron density parallel to the surface layer is observed between the defects. The results of the analysis of supercell size effects are similar to those for pure vacancies. However, in the case of surface U-vacancy and O atom, a larger concentration of electron density was observed between the O atom and neighbouring N atoms in the sub-surface layer in a comparison to the pure N-vacancy. Thus, the effect of slab thickness may not be ignored here, too.

In Fig. 12, the total and projected densities of states are shown for the 7-layer defective UN(001) surface with the O atom incorporated into the N-vacancy. The system remains conducting throughout all the calculations with the significant contribution from the U(5*f*) states at the Fermi level similar to perfect UN(001) slab. The appearance of specific O(2*p*) band with the energy peak at –6 eV is observed. a noticeable shift of the O(2*p*) band (by about -1.0 eV) allows one to distinguish the surface layer from the internal layers, when comparing the DOS for the O atoms incorporated into the N-vacancies.

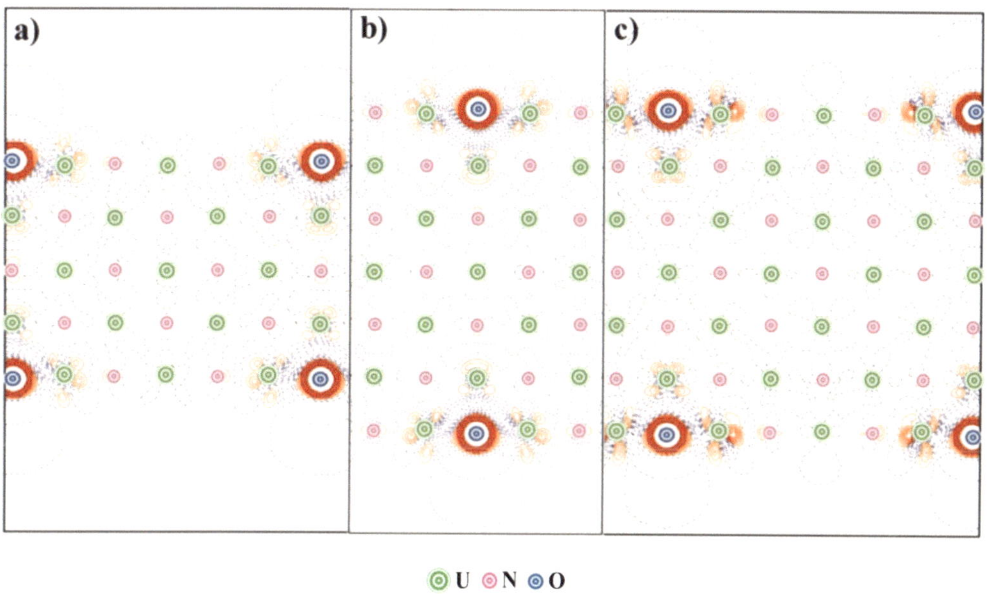


Fig. 11. The 2D sections of the electron charge density re-distributions $\Delta\rho(\mathbf{r})$ around the O atoms incorporated into the surface N-vacancies of the 5- and 7-layer UN(001) slabs with 2×2 and 3×3 supercell extensions. $\Delta\rho(\mathbf{r})$ are defined as the total electron density of the O-containing defected surface minus a superposition of the electron densities of the surface containing the N vacancies and the O atom in the regular positions on the surface. a) 3×3 periodicity of the O atoms upon the five-layer slab, b) 2×2 periodicity of the O atoms upon the seven-layer slab, c) 3×3 periodicity of the O atoms upon the 7-layer slab. Other details are given in caption of Fig. 9.

Moreover, in the case of surface layer, this band considerably overlaps with the N(2*p*) band, partly mixed with the U(5*f*) states (similar effects occur with the $O_2$ molecule atop the surface U atom [6]). In contrast, the O(2*p*) band remains quasi-isolated from the other bands (analogously to the O atom incorporated into the N-vacancy in the UN bulk [58]). Position of the N(2*p*) band is insensitive to presence of O atoms and lies within energy range of -6 and -1 eV.

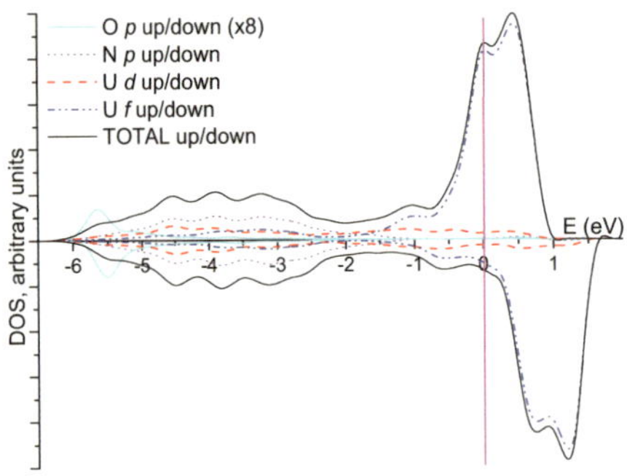


a)

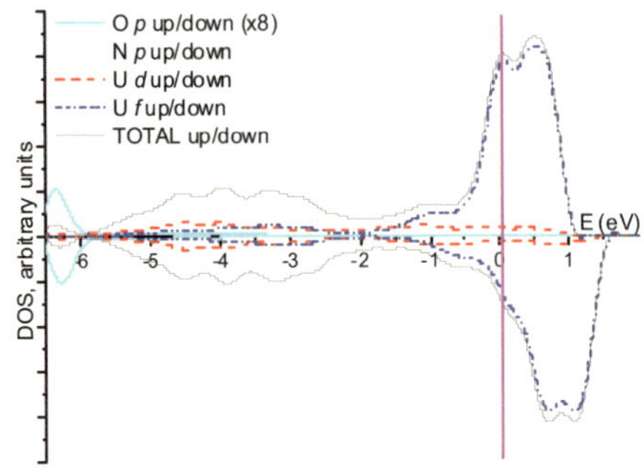


b)

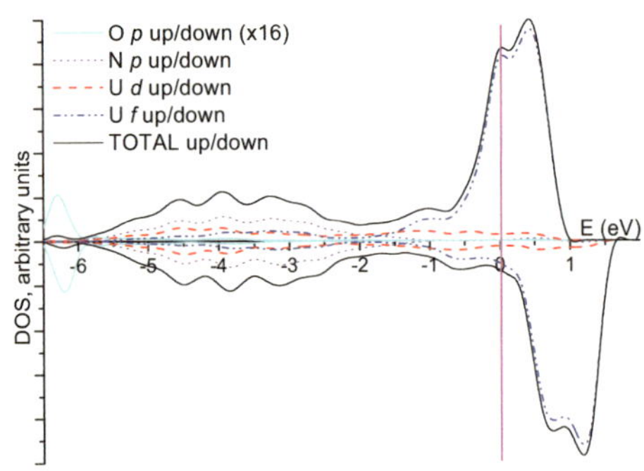


c)

Fig. 12. The total and projected DOS for three positions of O atoms incorporated into the N-vacancies of the 7-layer UN(001) slab with the 3x3 supercell: a) surface layer b) sub-surface layer, c) central layer. The O(2*p*) peaks were normalized to the same value, i.e., these have been multiplied by a factor of 8 and 16 for the vacancies in the surfrace (sub-surface) and central layers, respectively (see Figure labels). A convolution of individual energy levels was plotted using the Gaussian functions with a half-width of 0.2 eV.

### 8.4 Comparison of the UN (001) and (110) surfaces

Table 11 compares $E_S$ and $E_I$ for the two surfaces as functions of the slab thickness and supercell size. One can see that the UN(110) surface is characterized by more negative $E_S$ even though the difference between the solution energies is ~0.3 eV only. On the other hand, $E_I$ is charaterized by an opposite trend, suggesting more negative values for the (001) surface. Moreover, the difference between $E_I$ fort the two surfaces approaches 0.4 eV. Such results demonstrate importance of the $E_I$ calculations, as the role of different surfaces may also change with the temperature. Nevertheless, we clearly see similar trends for both of the surfaces (Table 11).

| Model | $E_I$ | $E_S$ | $q_{eff}$ | $E_I$ | $E_S$ | $q_{eff}$ |
|---|---|---|---|---|---|---|
| | (001) surface | | | (110) surface | | |
| 5, 2×2 | -6.17 | -2.47 | -1.36 | -5.85 | -2.78 | -1.27 |
| 7, 2×2 | -6.18 | -2.48 | -1.36 | -5.82 | -2.79 | -1.29 |
| 9, 2×2 | -6.19 | -2.48 | -1.36 | -5.82 | -2.78 | -1.29 |
| 11, 2×2 | -6.20 | -2.48 | -1.35 | -5.82 | -2.79 | -1.29 |
| 7, 3×3 | -6.13 | -2.48 | -1.36 | -5.75 | -2.78 | -1.28 |

Table 11. Incorporation ($E_I$) and solution ($E_S$) energies, effective charge of O atoms (q in $e^-$) for O incorporated into the N-vacancy of the UN (001) and (110) surfaces. The reference states were the same as in Table 10.

## 9. Conclusions

Based on the results of calculations discussed above the following stages of oxygen interaction with the UN surfaces were indentified to explain its easy oxidation: (*i*) chemisorption of molecular oxygen, (*ii*) spontaneous breaking of the $O_2$ chemical bond after molecular adsorption, (*iii*) location of the two newly formed O adatoms atop the adjacent surface U atoms, (*iv*) high mobility of adsorbed O atoms along the surface, (*v*) low-barrier incorporation of O into N-vacancies, *(vi)* stabilization of O atom inside the N-vacancy, (*vii*) further incorporation of O in pre-existed sub-surface N-vacancies as a result of inter-layer diffusion.

## 10. Acknowledgments

This study was partly supported by the EC FP7 F-BRIDGE project, ERAF project No. 2010/0204/2DP/2.1.1.2.0/10/APIA/VIAA/010, and ESF project No. 2009/0216/1DP/1.1.1.2.0/09/APIA/VIAA/044. Authors are indebted to P. van Uffelen, R.A. Evarestov, R. Devanathan, M. Freyss, E. Heifets, V. Kashcheyevs, Yu. A. Mastrikov, and S. Piskunov for helpful discussions. The technical assistance of A. Gopejenko, A. Gusev and A. Kuzmin was the most valuable.

## 11. References

[1] Hj. Matzke, Science of Advanced LMFBR Fuel, North Holland, Amsterdam, 1986.

[2] The Nuclear Fuel Cycle. P.D. Wilson (Eds.), University Press, Oxford, 1996.

[3] H. Wiame, M. Centeno, S. Pacard, P. Bastian, and P. Grange, Thermal oxidation under oxygen of zirconium nitride studied by XPS, DRIFTS, TG-MS. - J. Eur. Ceram. Soc., 1998, 18, p. 1293-1299.

[4] M. Walter, Oxidation of inert matrices, JRC-ITU-TN-2005/35 (Research report).

[5] R.A. Evarestov, A.V. Bandura, M.V. Losev, E.A. Kotomin, Yu.F. Zhukovskii, and D. Bocharov, A first principles DFT study in UN bulk and (001) surface: Comparative LCAO and PW calculations. - J. Comput. Chem., 2008, 29, p. 2079-2087.

[6] Yu.F. Zhukovskii, D. Bocharov, E.A. Kotomin, R.A. Evarestov, and A.V. Bandura, First principles calculations of oxygen adsorption on the UN(001) surface. - Surf. Sci., 2009, 603, p. 50-53.
[7] Yu.F. Zhukovskii, D. Bocharov, and E.A. Kotomin, Chemisorption of a molecular oxygen on the UN (001) surface: *ab initio* calculations. - J. Nucl. Mater., 2009, 393, p. 504-507.
[8] D. Bocharov, D. Gryaznov, Yu.F. Zhukovskii, and E.A. Kotomin, DFT calculations of point defects on UN(001) surface. - Surf. Sci., 2011, 605, p. 396-400.
[9] D. Bocharov, D. Gryaznov, Yu.F. Zhukovskii, E.A. Kotomin, *Ab initio* modeling of oxygen impurity atom incorporation into uranium mononitride surface and subsurface vacancies. - J. Nucl. Mater., 2011, 416, p.200-204.
[10] D. Bocharov, Yu.F. Zhukovskii, D. Gryaznov, and E.A. Kotomin, *Ab initio* simulation on oxygen adatom migration upon UN (001) surface. - Surf. Sci., *submitted.*
[11] N. Curry, An investigation of the magnetic structure of uranium nitride by neutron diffraction. - Proc. Phys. Soc., 1965, 86, p. 1193-1198.
[12] T. Muromura and H. Tagawa, Lattice parameter of uranium mononitride. - J. Nucl. Mater., 1979, 79, p. 264-266.
[13] P.E. Evans and T.J. Davies, Uranium nitrides. - J. Nucl. Mater., 1963, 10, p. 43-55.
[14] N.-T.H. Kim-Ngan, A.G. Balogh, L. Havela, and T. Gouder, Ion beam mixing in uranium nitride thin films studied by Rutherford Backscattering Spectroscopy. - Nucl. Instr. Meth. Phys. Res. B, 2010, 268, p. 1875–1879.
[15] P.R. Norton, R.L. Tapping, D.K. Creber, and W.J.L. Buyers, Nature of the 5*f* electrons in uranium nitride: A photoelectron spectroscopic study of UN, U, $UO_2$, ThN, and Th. - Phys. Rev. B, 1980, 21, p. 2572-2577.
[16] T. Ito, H. Kumigashira, S. Souma, T. Tahakashi, and T. Suzuki, High-resolution angle-resolved photoemission study of UN and USb; Dual character of 5*f* electrons. - J. Magn. Magn. Mater., 2001, 226-230, p. 68-69.
[17] M. Marutzky, U. Barkow, J. Schoenes, and R. Troć, Optical and magneto-optical properties of single crystalline uranium nitride. - J. Magn. Magn. Mater., 2006, 299, p. 225–230.
[18] M. Paljević and Z. Despotović, Oxidation of uranium mononitride. - J. Nucl. Mater., 1975, 57, p. 253-257.
[19] Y. Arai, M. Morihira, and T. Ohmichi, The effect of oxygen impurity on the characteristics of uranium and uranium-plutonium mixed nitride fuels. - J. Nucl. Mater., 1993, 202, p. 70-78.
[20] L. Black, F. Miserque, T. Gouder, L. Havela, J. Rebizant, and F. Wastin, Preparation and photoelectron spectroscopy study of $UN_x$ thin films. - J. Alloys Comp., 2001, 315, p. 36–41.
[21] G.W. Chinthaka Silva, Ch.B. Yeamans, L. Ma, G.S. Cerefice, K.R. Czerwinski, and A.P. Sattelberger, Microscopic characterization of uranium nitrides synthesized by oxidative ammonolysis of uranium tetrafluoride. - Chem. Mat., 2008, 20, p. 3076-3084.
[22] M. Eckle, and T. Gouder, Photoemission study of $UN_xO_y$ and $UC_xO_y$ in thin films. - J. Alloys Comp., 2004, 374, p. 261–264.
[23] S. Sunder and N.H. Miller, XPS and XRD studies of corrosion of uranium nitride by water. - J. Alloys Comp., 1998, 271–273, p. 568–572.

[24] B. Reihl, G. Hollinger, and F.J. Himpsel, Itinerant 5*f*-electron antiferromagnetism in uranium nitride: A temperature-dependent angle-resolved photoemission study. - Phys. Rev. B, 1983, 28, p. 1490–1494.
[25] P. Weinberger, C.P. Mallett, R. Podloucky, and A. Neckel, The electronic structure of HfN, TaN and UN. - J. Phys. C: Solid St. Phys., 13, 1980, p. 173-187.
[26] M.S. Brooks and D. Glötzel, Some aspects of the electronic structure of uranium pnictides and chalcogenides. - Physica B, 1980, 102, p. 51-58.
[27] M.S. Brooks, Electronic structure of NaCl-type compounds of the light actinides. I. UN, UC, and UO. - J. Phys. F: Met. Phys., 1984, 14, 639-652.
[28] D. Sedmidubsky, R.J.M. Konings, and P. Novak, Calculation of enthalpies of formation of actinide nitrides. - J. Nucl. Mater., 2005, 344, p. 40–44.
[29] G.K. Johnson and E.H.P. Cordfunke, The enthalpies of formation of uranium mononitride and α- and β-uranium sesquinitride by fluorine bomb calorimetry. - J. Chem. Thermodyn., 1981, 13, p. 273-282.
[30] R. Atta-Fynn and A.K. Ray, Density functional study of the actinide nitrides. - Phys. Rev. B, 2007, 76, 115101 (p. 1-12).
[31] P.F. Weck, E. Kim, N. Balakrishnan, F. Poineau, C.B. Yeamans, and K.R. Czerwinski, First-principles study of single-crystal uranium mono- and dinitride. - Chem. Phys. Lett., 2007, 443, p. 82–86.
[32] Y. Lu, B.-T. Wang, R.-W. Li, H. Shi, and P. Zhang, Structural, electronic, and thermodynamic properties of UN: Systematic density functional calculations. - J. Nucl. Mater., 2010, 406, p. 218–222.
[33] B. Dorado, B. Amadon, M. Freyss, and M. Bertolus, DFT+*U* calculations of the ground state and metastable states of uranium dioxide. - Phys. Rev. B, 2010, 79, 235125 (p. 1-8)
[34] B. Dorado, G. Jomard, M. Freyss, and M. Bertolus, Stability of oxygen point defects in $UO_2$ by first-principles DFT+*U* calculations: Occupation matrix control and Jahn-Teller distortion. - Phys. Rev. B, 2010, 82, 035114 (p. 1-11).
[35] D. Gryaznov, E. Heifets and E.A. Kotomin, *Ab initio* DFT+*U* study of He atom incorporation into $UO_2$ crystals. - Phys. Chem. & Chem. Phys., 2009, 11, p. 7241-7247.
[36] D. Rafaja, L. Havela, R. Kuel, F. Wastin, E. Colineau, and T. Gouder, Real structure and magnetic properties of UN thin films. – 2005, 386, p. 87-95.
[37] E.A. Kotomin, Yu.A. Mastrikov, Yu.F. Zhukovskii, P. Van Uffelen, and V.V. Rondinella, First-principles modelling of defects in advanced nuclear fuels. - Phys. Stat. Sol. (c), 2007, 4, p. 1193-1196.
[38] E.A. Kotomin, R. W. Grimes, Yu. A. Mastrikov, and N.J. Ashley, Atomic scale DFT simulations of point defects in uranium nitride. - J. Phys.: Cond. Mat, 2007, 19, 106208 (p. 1-9).
[39] D. Gryaznov, E. Heifets, D. Sedmidubsky, Density functional theory calculations on magnetic properties of actinide compounds, Phys. Chem. & Chem. Phys., 2010, 12, p. 12273-12278.
[40] E.A. Kotomin, D. Gryaznov, R.W. Grimes, D. Parfitt, Yu.F. Zhukovskii, Yu.A. Mastrikov, P. Van Uffelen, V.V. Rondinella, and R.J.M. Konings, First-principles modelling of radiation defects in advanced nuclear fuels. - Nucl. Instr. Meth. Phys. Res. B, 2008, 266, p. 2671–2675.

[41] G. Kresse and J. Furthmüller, *VASP* the Guide, University of Vienna, 2009;http://cms.mpi.univie.ac.at/vasp/
[42] J. Hafner, *Ab initio* simulations of materials using VASP: Density-Functional Theory and beyond. - J. Comput. Chem., 2008, 29, p. 2044–2078.
[43] G. Kresse and J. Furthmüller, Efficient iterative schemes for *ab initio* total-energy calculations using a plane-wave basis set. - Phys. Rev. B, 1996. 54, p. 11169-11186.
[44] G. Kresse and D. Joubert, From ultrasoft pseudopotentials to the projector augmented-wave method. - Phys. Rev. B, 1999, 59, p. 1758-1775.
[45] H.J. Monkhorst and J.D. Pack, Special points for Brillouin-zone integrations. - Phys. Rev. B, 1976, 13, p. 5188-5192.
[46] P.W. Tasker, The stability of ionic crystal surfaces. - J. Phys. C: Solid State Phys., 1979, 12, p. 4977-4984.
[47] J. Akella, S. Weir, J. M. Wills, and P. Söderlind, Structural stability in uranium. - J. Phys.: Condens. Matter, 1997, 9, L549 (p. 1-7).
[48] C.G. Van de Walle and J. Neugebauer, First-principles calculations for defects and impurities: Applications to III-nitrides. - J. Appl. Phys., 2004, 95, p. 3851-3879.
[49] P. Söderlind, First-principles elastic and structural properties of uranium metal. - Phys. Rev. B, 2002, 66. 085113 (p. 1-7).
[50] B. Dorado, M. Freyss, and G. Martin, GGA+*U* study of the incorporation of iodine in uranium dioxide. - Eur. Phys. J. B, 2009, 69, p. 203-210.
[51] M. Freyss, First-principles study of uranium carbide: Accommodation of point defects and of helium, xenon, and oxygen impurities. - Phys. Rev. B, 2010, 81, 014101 (p. 1-16).
[52] D.R. Lide (ed.), CRC Handbook of Chemistry and Physics, 88$^{th}$ Edition, CRC Press (2007-2008).
[53] M. Iwasawa, Y. Chen, Y. Kaneta, T. Ohnuma, H. Y. Geng, and M. Kinoshita, First-principles calculation of point defects in uranium dioxide. - *Mat. Trans*, 2006, 47, p. 2651- 2657.
[54] D. Gryaznov, M. Finnis, HPC-Europa2, Science and Supercomuting in Europe, Research Highlights, 2010, p. 127 http://www.hpc-europa.org/files/SSCinEurope/CD2010/contents/127-material-Denis.pdf
[55] Yu.F. Zhukovskii, P.W.M. Jacobs, and M. Causà, On the mechanism of the interaction between oxygen and close-packed single-crystal aluminum surfaces. - J. Phys. Chem. Solids, 2003, 64, p. 1317-1331.
[56] R. Weast, CRC Handbook of Chemistry and Physics. CRC Press Inc., Boca Baton (FL), 1985.
[57] S. Piskunov, Yu.F. Zhukovskii, E.A. Kotomin, E. Heifets, and D. E. Ellis, Adsorption of atomic and molecular oxygen on the $SrTiO_3$(001) surfaces: Predictions by means of hybrid density functional calculations. - MRS Proc., 2006, 894, LL08-05 (p. 1-6).
[58] E. A. Kotomin and Yu.A. Mastrikov, First-principles modelling of oxygen impurities in UN nuclear fuels. - J. Nucl. Mater., 2008, 377, p. 492-495.
[59] R. W. Grimes and C.R.A. Catlow, The stability of fission products in uranium dioxide. – Phil. Trans. Roy. Soc. A, 1991, 335, p. 609-634.

# 6

# Nuclear Accidents in Fukushima, Japan, and Exploration of Effective Decontaminant for the $^{137}$Cs-Contaminated Soils

Hajime Iwata*, Hiroyuki Shiotsu*,
Makoto Kaneko* and Satoshi Utsunomiya*,**
*Department of Chemistry, Kyushu University, Hakozaki, Higashi-ku, Fukuoka*
*Japan*

## 1. Introduction

### 1.1 A brief summary of the accidents at the Fukushima Dai-ichi nuclear power plant and the release of radionuclides to the environment

Nuclear accident at the Fukushima Dai-ichi nuclear power plant (FDNPP), which is located 230 km north of downtown Tokyo, was the most recent tragedy in the nuclear society. The total release of radioactivity was estimated to be $6.3 \times 10^{17}$ Bq, which is approximately one-tenth of the total radioactivity released from the Chernobyl ($5.2 \times 10^{18}$ Bq) (TEPCO, 2011). The International Atomic Energy Agency (IAEA) has recently announced the International Nuclear and Radiological Event Scale (INRES) of the Fukushima accident to be level 7.

There are six nuclear reactors in the FDNPP (Fig. 1). These boiling water reactors (BWRs) generated 460 (reactor 1), 784 (reactor 2-5), and 1100 (reactor 6) mega watt. Thirty-two out of 548 fuel rods were mixed oxide fuel (MOX) in the reactor 3, of which the Pu concentration in each fuel pellet contains up to 10 wt% of plutonium.

When the earthquake with the magnitude of 9.0 hit the east Japan on March 11th, the reactors 1-3 were in operation. These reactors immediately shut down by inserting the control rods and the nuclear reaction stopped in the reactors 1-3. However, the catastrophic Tsunami with the water level unexpectedly as high as ~15 m caused serious damage in the power plant, including the power outage and the following cut-off in the emergency power supply (Narabayashi & Sugiyama, 2011). The power plant was facilitated with a breakwater based on the expected Tsunami level as high as 5.7 m, which was far not enough for the Tsunami this time. The power outage caused serious problems, because the cut-off in the emergency electric system stopped the watering systems cooling down the nuclear reactor cores. The loss of water supply and continuous escape of steamed water to the pressure suppression system lowered the level of the cooling water, resulting in full or partial exposition of the fuel rods. The temperature of nuclear fuels increased upon exposing out

* All authors contributed equally
** Author for correspondence

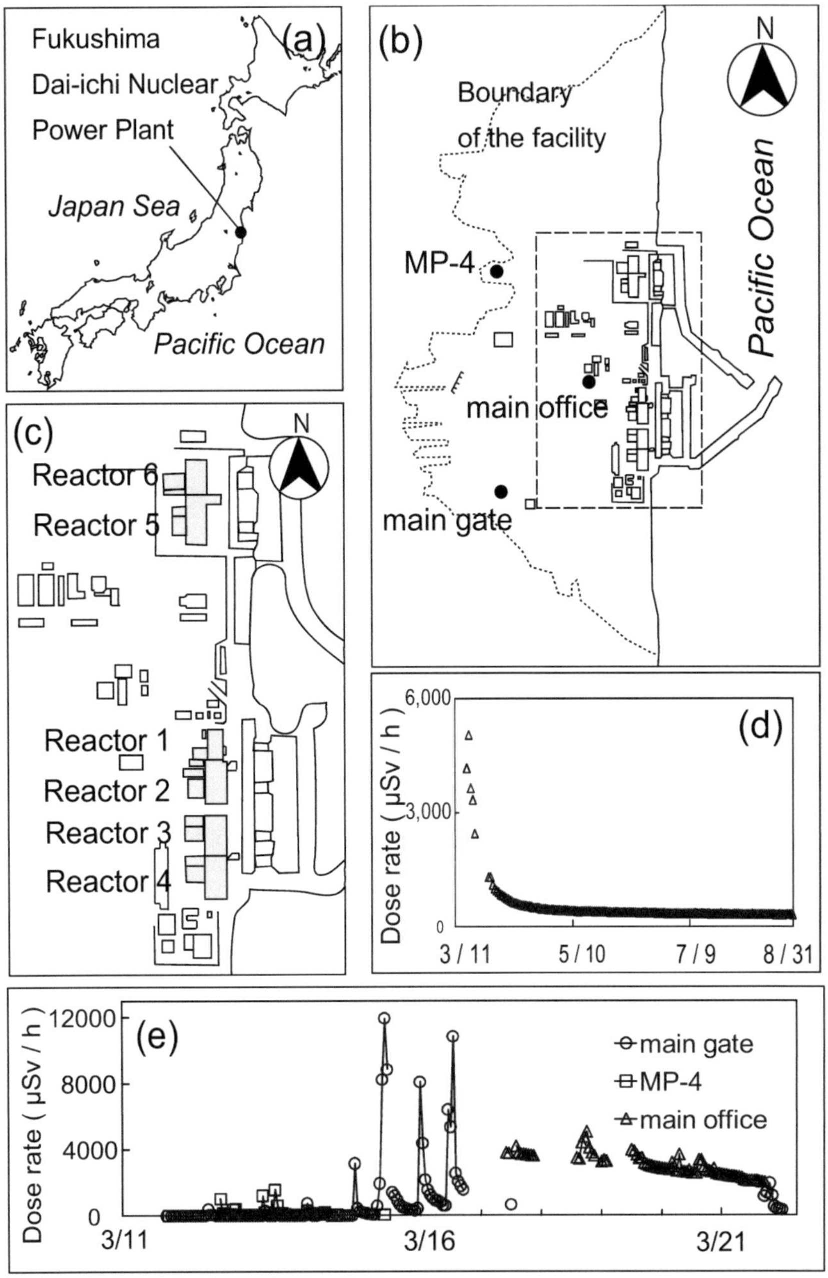


Fig. 1. The Fukushima Dai-ichi Nuclear Power Plant (FDNPP) and radioactivities in the facility reported by Tokyo Electric Power Company (TEPCO). (a) The map of Japan showing the location of the FDNPP. (b) The map of the FDNPP with the location of monitoring posts. (c) A magnified illustration showing the position of the reactors 1-6. (d) Time-dependent radioactivity monitored at the main office reported by TEPCO. (e) Time-dependent radioactivities during the first 10 days after the earthquake detected at three monitoring posts. Data from the TEPCO web site (TEPCO, 2011).

of water, and at the elevated temperature, Zr metal alloy that clads the fuel reacted with water producing $H_2$ as the following expression:

$$Zr + 2H_2O \rightarrow ZrO_2 + 2H_2$$

A part or a whole body of the fuel melted down due to the temperature increase derived from the decay heat beyond the melting temperature of the $UO_2$ without cooling water. The melted core broke through the pressure vessel and fission products leaked into the containment vessel. Subsequently, these fission products were released from the containment vessel due to the high inside pressure as well as $H_2$ gas. The hydrogen gas accumulated in the reactor building. As a result, $H_2$ explosion occurred at the operation floor in the reactor 1, 3, 4 and possibly in the pressure control unit of the reactor 2.

According to the calculation by Naito (2011), the amount of the produced $H_2$ gas was estimated to be ~500 kg and the pressure inside the reactor 1 building reached ~5 atm at the time of "deflagration". In the case of the reactor 3, the pressure was estimated as high as ~60 atm during the "detonation". The reactor 4, which hosted 1,331 spent nuclear fuels in the storage pool, was in shut down mode for the regular check and maintenance at the time of earthquake. However, the excess $H_2$ gas generated in the reactor 3 in-flowed through the pipelines connecting between the reactors 3 and 4.

During the series of those initial crises, various efforts were made to cool down the reactor cores of the reactors 1-3 by using seawater and subsequently freshwater. This water injection generated a large amount of radioactive waste waters, of which the total amounts of the injected waters until May 31 were 13,630 $m^3$, 20,991 $m^3$, and 20,625 $m^3$ for the reactors 1, 2, and 3, respectively (TEPCO, 2011). The series of major events during the initial stage of the accident at the FDNPP is summarized in Table 1.

There are two serious issues raised by the series of accidents. One is the release of the large amount of highly radioactive water to the near-field environment, of which the solvent is a mixture of freshwater and seawater. 10,393 $m^3$ of radioactive water with the total radioactivity of $1.5 \times 10^{15}$ Bq was released out to the ocean, while 19,770 $m^3$ of the waste water was stored and isolated in the main processing building and the high temperature calcinator building (TEPCO, 2011). A new circulation system of cooling water was installed and started operating on June 27, which is equipped with a series of filtering and ion-exchanging columns.

Although a large amount of radionuclides was locally released to the Fukushima Pacific coast, the results of monitoring radioactivity revealed the minimum contamination in seawater owing to dilution by almost infinite volume of seawater. As far as the water purifying system operates, the contamination will keep decreasing. The high-level radioactive waters stored in the main processing building and the high temperature calcinator building, of which the total amount is 19,770 $m^3$, have been treated by means of zeolite-based materials adsorbing radionuclides, namely Cs radioisotopes. On the other hand, immediately after the accident, a number of experiments were systematically conducted by the research groups at some Japanese universities to evaluate the adsorption coefficients of $I^-$ and $IO_3^-$ to a large set of adsorbents in the mixed waters of seawater and freshwater; that is, with high ionic strength. The highest adsorption coefficient ($K_d$ ~$10^5$) was found in the use of cobalt ferrocyanide in seawater.

| Reactor # | 1 | 2 | 3 | 4 | 5 | 6 |
|---|---|---|---|---|---|---|
| Status | Under operation | | | Under routine maintenance | | |
| March 11 | Mw9.0 earthquake (14:46) | | | | | |
| | Automatically shut down | | | | | |
| | TSUNAMI (15:27) | | | | | |
| | Lost all electrical power supply. (15:37) Emergency Core Cooling System was shut down. (16:36) | Lost all electrical power supply.(15:41) Emergency Core Cooling System was shut down. (16:36) | Lost all electrical power supply. (15:38) | Lost all electrical power supply. (15:38) | Safely stopped | |
| March 12 | The water level in the reactor reached the top of the fluel. (12:55) Malfunction of pressure suppression system Hydrogen detonation. (15:36) | Malfunction of pressure suppression system. (10:58) | | | | |
| March 13 | | | The water level in the reactor reached the top of the fluel.(4:15) Emergency Core Cooling System was shut down. (5:10) | | | |
| March 14 | | The reactor core isolation cooling system for the reactor stop. (13:25) The water level in the reactor reached the top of the fluel. (18:00) | Abnormal rise in pressure of nuclear containment. Hydrogen detonation. (11:00) | Rise in water temperature of spent fuel pool. (14:30) | | |
| March 15 | | Explosion in the "pressure suppression room". (16:00) | | A fire beaked out. (6:00) | | |
| March 20 | | | The pressure in nuclear containment vessel was rising.(11:00) | | | |

Table 1. A summary of the major events occurred in the nuclear reactors at the FDNPP during the initial stage of the series of accidents.

The other issue is contamination of surface soils in the vicinity of the FDNPP. Figure 1d shows the released radioactivities to ambient atmosphere for the period till the end of August 2011, indicating dramatic decrease in the radioactivity by June; approximately 2 millionth from the time of accident. As of August in 2011, TEPCO is constructing the shield to cover the collapsed top part of the reactor buildings to completely shut off the release of radionuclides, and further contamination will be stopped. However, a large area of the Fukushima prefecture has been already contaminated with $^{137}$Cs ($T_{1/2}$ = 30.07 y), $^{134}$Cs ($T_{1/2}$ = 2.062 y), $^{131}$I ($T_{1/2}$ = 8.02 d), and minor $^{90}$Sr ($T_{1/2}$ = 29.1 y) , as high as several hundreds mSv/h near the power plant. Other fissiogenic elements released into the surrounding environment are summerized in Table 2. The total volume of the contaminated soils was estimated to be 28,785,000 m$^3$ (MEXT, 2011).

| |
|---|
| $^{131}$I(<1,000,000), $^{137}$Cs(<430,000), $^{134}$Cs(<370,000), $^{129m}$Te(<180,000), $^{129}$Te(<50,000), $^{140}$La(<1,900), $^{110m}$Ag(<1,600), $^{140}$Ba(<1,600), $^{89}$Sr(<1,500), $^{136}$Cs(<1,000), $^{95}$Nb(<530), $^{90}$Sr(<250), $^{234}$U(<18.0), $^{238}$U(<17.0), $^{235}$U(<0.82), $^{239+240}$Pu(<0.05), $^{242}$Cm(<0.032), $^{241}$Am(<0.028) |

Table 2. List of the released fissiogenic elements detected in soils in the vicinity of the FDNPP. The values in parenthesis stand for the maximum concentration (Bq/kg) reported to date, which were compiled from the TEPCO database (TEPCO, 2011).

For predicting the distribution and mobility of the released radionuclides, the System for Prediction of Environmental Emergency Dose Information (SPEEDI) was formerly developed and applied to the case of Chernobyl accident (Chino et al., 1986). The SPEEDI was also utilized for the case of the Fukushima accident and the prediction of radioactivity was opened to public (Chino et al., 2011). On the other hand, owing to the well-established network of monitoring post and the rigorous sampling campaign conducted after the accident, the dose rate was measured and the dose accumulation was predicted for the one year duration since the accident.

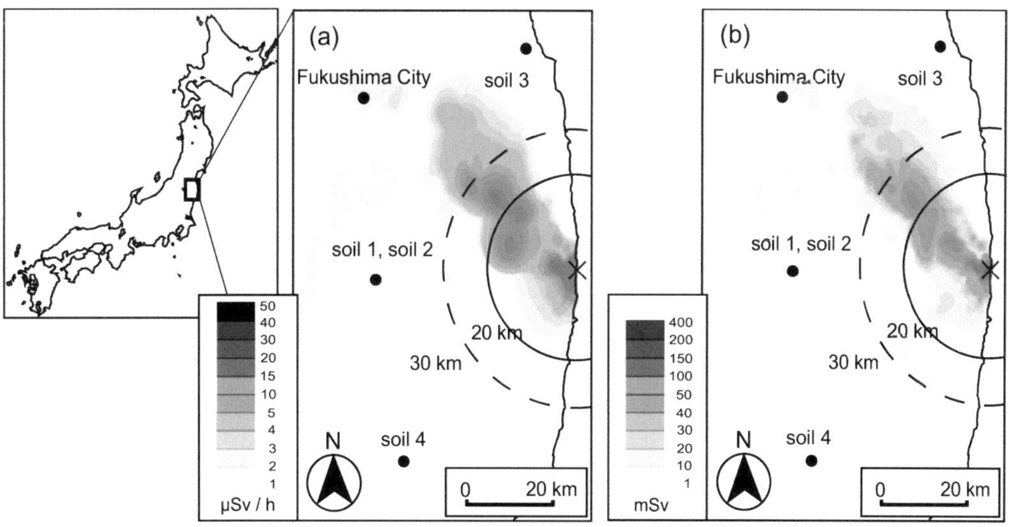


Fig. 2. The maps of the radioactivity distribution measured 1 m above the ground. Data from Ministry of Education, Culture, Sports, Science and Technology, Japan (MEXT, 2011). (a) Measured radioactivity on 8/11/2011. (b) Prediction of the accumulated radioactivity for one year period; 3/11/2011-3/11/2012.

Figures 2a and b show the map of the radioactivity on 8/11/2011 near ground level (a) and the contour map of the accumulated dose estimated for the period from March 11, 2011 to March 11, 2012 (b). The highly contaminated area is characteristically elongated towards the north-west covering Futaba, Namie town and Iidate village. This biased distribution is attributed to the wind direction during the explosive events. In the area near the FDNPP, the wind was often directed to the northwest because of the sea-breeze in day time and the topological characteristics (Yamazawa and Hirao, 2011). Especially, the amount of released radionuclides on March 15$^{th}$ was estimated to be one to two orders of magnitude greater than that in the other days and the wind was directed toward the northwest (Yamazawa and Hirao, 2011). Because of the biased wind direction, $^{134}$Cs, $^{137}$Cs and $^{131}$I in the surface soils at Komiya, Iidate village, were detected to be ~140,000 Bq/kg, ~170,000 Bq/kg, and ~1,100 Bq/kg, respectively, as of June 5, 2011, where is even 35 km distant from the FDNPP (MEXT, 2011). The initially reported values analyzed for the soil from Tsushima, Namie town, located 30 km distant from the FDNPP revealed much higher dose rate; 282,000 Bq/kg for $^{134}$Cs, 290,000 Bq/kg for $^{137}$Cs, and 710,000 Bq/kg for $^{131}$I on March 30, 2011 (NSCJ, 2011).

## 2. Exploration of efficient decontamination protocol

### 2.1 Background of experiments

Because the large-scale contamination in the vicinity of the FDNPP is mainly derived from $^{137}$Cs, the efficient protocol to decontaminate Cs that already adsorbed to the soils is highly demanded in Fukushima prefecture and even in near Tokyo.

Adsorption/desorption phenomena of Cs to/from various geological media have been extensively investigated in the previous studies to evaluate the capacity of Cs immobilization by natural or artificial buffer materials (Comans & Hockley, 1992; Khan et al. 1994; Hsu & Chang, 1994; Seaman et al., 2001; Hassan 2005; Mon et al., 2005; Bellenger & Staunton, 2008; Rajec & Domianova, 2008; Wang et al., 2010; Yildiz et al. 2011) or to understand the migration of Cs in the specific contaminated sites (Zachara et al., 2002; Liu et al., 2003; Steefel et al., 2003; Missana et al., 2004; Bouzidi et al., 2010).

In the present study, Cs adsorption and desorption experiments have been carried out in laboratory using non-contaminated soils collected from Fukushima. It is noted that the goal of this chapter is not to discuss either kinetics or mechanism of Cs adsorption/desorption process. We demonstrate detailed characterization of soil materials utilizing a variety of high resolution techniques including electron microscopy to understand; i) the type of clay minerals that strongly bound to Cs, ii) quantity of adsorbed and desorbed Cs. We also test a series of desorption experiments exploring the most effective protocol to remove Cs from the contaminated soils.

### 2.2 Experimental methods

#### 2.2.1 Soil samples

For adsorption and desorption experiments in the present study, we have collected four surface soils from different locations in Fukushima prefecture (plotted in Fig. 2) in June 2011. These sites have not been contaminated, of which the radioactivities are as low as the background level.

The surface area of the soil samples was analyzed by a Brunauer Emmett Teller (BET) method using Quantachrome AUTOSORB-1 with $N_2$ gas. Each sample was heated at 120 °C for 1.5 h to remove adsorbed water prior to the mesurement.

Cation exchange capacity (CEC) was also determined for all samples. About 2.0 g soil sample was placed in 30 ml of 0.1 M $BaCl_2$(aq) and intermittently shaken for 1 h, followed by centrifugation at 6000 rpm for 10 min. This procedure was repeated for three times to completely replace exchangeable cations with Ba. Subsequently, the solid phase was contacted with 30 ml of 0.0025 M $BaCl_2$(aq) for 14 h. Then, the treated soil was contacted with 0.02 M $MgSO_4$ (aq) for 28 h and centrifuged. The supernatant was filtered by 0.20 μm syringe filter and the Mg concentration was measured by an atomic absorption spectrometry (AA, Shimadzu AA-6300). The CEC can be calculated using the following expression:

$$C_1' = C_1 \times \frac{(a + m_2 - m_1)}{a}$$

$$CEC = \frac{a\left(C_1' - C_2\right)}{m_1} \times 100$$

$C_1$ : initial Mg concentration (mol/l)
$C_1$ ‘: corrected initial Mg concentration (mol/l)
$C_2$ : final Mg concentration (mol/l)
$m_1$ : dry soil mass (g)
$m_2$ : wet soil mass (g)
$a$ : 0.02 mol/l $MgSO_4$ volume (ml)
Fundamental information and properties of the soils used in the experiments are summarized in Table 3.

| Sample | Soil 1 | Soil 2 | Soil 3 | Soil 4 |
|---|---|---|---|---|
| Location | Miharu town<br>N37°24'30.8",<br>E140°29'59.7" | Miharu town<br>N37°23'37.5",<br>E140°30'14.6" | Shinchi town<br>N37°51'31.8",<br>E140°54'45.2" | Furudono town<br>N37°05'14.2",<br>E140°33'40.8" |
| Occurrence | Temple yard | Garden soil | Park | Elementary school play yard |
| Depth | 0-5 cm | 0-5 cm | 10-15 cm | 0-5 cm |
| Surface area ($m^2$/g) | 2.32 | 5.15 | 27.8 | 7.82 |
| CEC (cmol/kg) | 1.24 ± 0.18 | 1.77 ± 0.15 | 4.53 ± 0.02 | 2.49 ± 0.28 |

Table 3. Basic information and properties of the four non-contaminated soils collected in Fukushima.

### 2.2 Adsorption / desorption batch experiments

Adsorption experiment was carried out in batch system. About 50 g of soil sample was contacted with 1 l of CsCl solution with three different Cs concentrations: 10, 1.0, and 0.1 mM. 3 ml of supernatant was collected at 1, 2, 4, 8, 24, 48 hours, 4, 7, 14, 21 and 28 days, and filtered with a 0.2 μm-pore-sized syringe filter. The filtrates were diluted to $10^3$-$10^5$ times with milli-Q water for the solution analysis.

After the adsorption experiment, the soil was separated with 0.2 μm pore-sized filter and air-dried at room temperature to prepare the starting materials of batch desorption experiment. Six extractants: deionized water (DW), 0.1 mol/l KCl, $NH_4Cl$, $MgCl_2$, acetic acid, and citric acid, were tested in the desorption experiment. 200 ml of the extractant was contacted with 5 g of the soil 1 or 3 that preliminarily adsorbed Cs in the 10 mM CsCl solution. During the desorption experiment, mixture of the soil and solution was continuously agitated. The 1 ml of suspension was collected and filtered through a 0.2 μm-pore-sized syringe filter at 1, 2, 4, 8, 24, and 48 hours, 4 and 7 days. The filtrates were diluted to $10^3$ times for the solution analysis.

### 2.3 Analytical methods

The Cs concentration in solution was analysed by an inductively coupled plasma mass spectrometry (ICP-MS, Agilent 7500c) in non gas mode with In internal standard. The calibration line was drawn with 0, 1, 10, 50, 100, 200 ppb and the detection limit of Cs was ~0.1 ppt.

The major mineral assemblage of the soils was determined by using a powder X-ray diffraction (XRD, RIGAKU MultiFlex) equipped with a Cu target and a reflected beam monochromator. The scan rage was $3^o$-$63^o$ with the scanning speed of $1^o min^{-1}$ of $2\theta$ and the step angle of $0.02^o$. In addition to the bulk samples, the levigated soil samples were measured to obtain more detailed information on clay minerals. About 15 g of the soil samples were suspended in DW, ultrasonicated for 5 min. and centrifuged at 1000 rpm for 10 min. The clay-rich portion was separated and air-dried at room temperature on a watch glass. The XRD was performed ranging $3^o$-$33^o$ at the scan speed of $1^o min^{-1}$ of $2\theta$ with the step angle of $0.02^o$.

Individual particle analysis was performed for the soil samples contacted with 10 mM CsCl solution by using a scanning electron microscopy (SEM, SHIMADZU SS-550) equipped with energy dispersive X-ray spectroscopy (EDX). Topology-sensitive imaging was conducted at 5 kV of the acceleration voltage and EDX analysis was completed at 25 kV. All samples were coated with carbon using a carbon coater (SANYU SC-701C) to make conductivity.

### 2.4 Results and discussion

#### 2.4.1 Soil characterization by XRD

The soil samples were first characterized by XRD analysis to determine the major mineral phases and the results are shown in Fig. 3. In all soil samples, the major mineral phases are quartz and feldspar. Amphibole was additionally detected in soil 1 and 2. Peaks of clay minerals were not clearly recognized in these measurements.

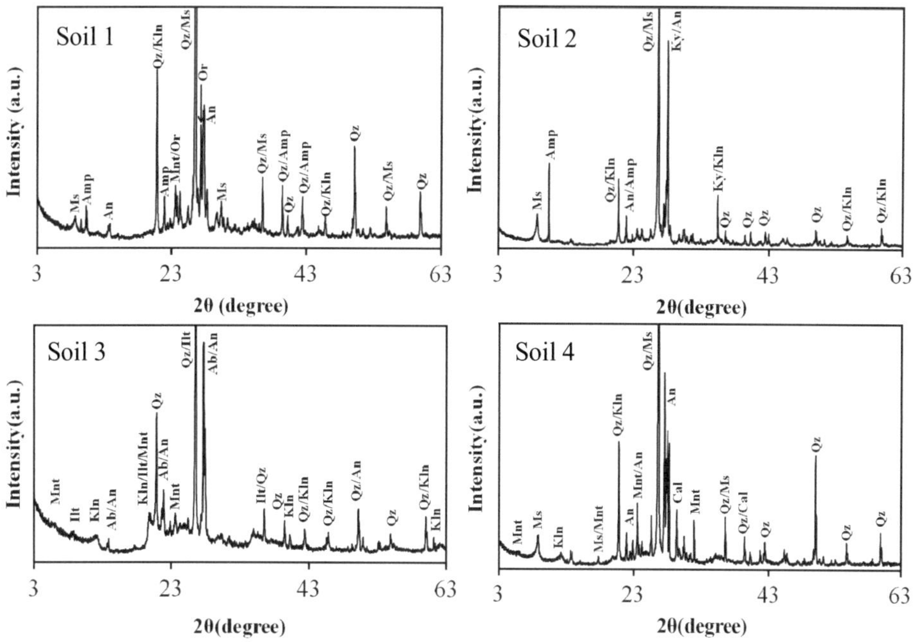


Fig. 3. The results of XRD analysis of the four soils from Fukushima showing the major mineral components. Ms: muscovite, Amp: amphibole, Qz: quartz, Mont: montmorillonite, An: anorthite, Or: orthoclase, Kln: kaolinite, Ilt: illite

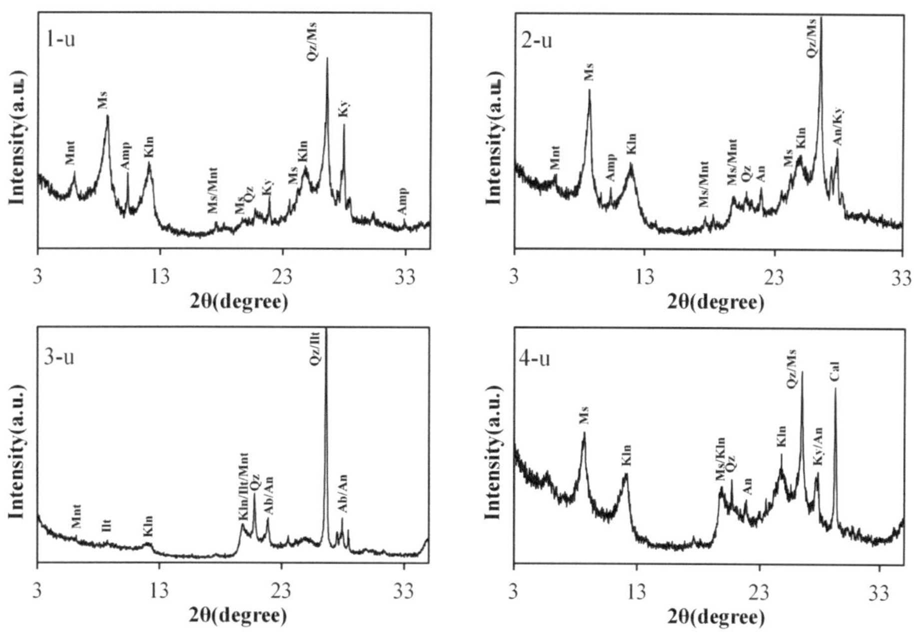


Fig. 4. The results of XRD analysis of the four Fukushima soils after ultrasonication and levigation. The scan range was limited to the lower angles in order to focus on the characterization of clay minerals.

Hence, these soils were subsequently levigated as described in the experimental method. The results are shown in Fig. 4. All profiles revealed peaks at ~7, ~10, and ~14 Å, which correspond to kaolinite, muscovite/illite, and montmorillonite/chlorite, respectively, although the relative intensities at these peak positions vary depending on the soil origion. The soil 3 contained the least amount of sheet silicate minerals. Ethylene glycol (EG) treatment (data not shown here) allowed to distinguish presence of chlorite and smectite, and the peak at ~14 Å remained in all soil samples, indicating that the major phase responsible for ~14 Å peak is chlorite.

### 2.4.2 Cesium adsorption to the Fukushima soils

Table 4 summarizes the time-dependent Cs concentration in the solutions during the adsorption experiments. The data were also plotted in Fig. 5. They reveal that the Cs concentration drastically decreased after contacting the soils. The Cs concentration reached the apparent equilibrium after 336 hours in most of the soils. The time to reach the apparent equilibrium was almost the same for all three initial Cs-concentrations: $10^{-1}$, $10^{0}$, and $10^{1}$ mM. The distribution ratio ($K_d$) between soil and solution was calculated by the following equation:

$$K_{\mathrm{d,adsorption}} = \frac{(C_i - C_f)}{C_f} \times \frac{W_l}{W_s}$$

$C_i$ : the initial Cs concentration (mol/l)
$C_f$ : the final Cs concentration (mol/l)
$W_l$ : the solution volume (l)
$W_s$ : the soil mass (kg)

The calculated $K_d$s are given in Table 4.

As seen in the table, the $K_d$ value increases as the initial concentration increases. The $K_d$ also showed the trend of soil 3 > soil 4 > soil 2 > soil 1 for the initial Cs concentration of 1 and 10 mM, while soil 4 revealed the highest $K_d$ at the $10^{-1}$ mM. The high $K_d$ in the soil 3 can be easily explained by its highest surface area and CEC. However, the explanation of the high $K_d$ in the soil 4 at the low Cs concentration requires the other mechanism besides the factors of surface area and CEC. This may be attributed to variation in the amount of two different sites for Cs adsorption; high affinity site and low affinity site in the mineral component. Indeed, Rajec et al. (1999) reported a positive correlation between total surface area and Cs sorption capacity, while inconsistent correlation between total surface area and distribution coefficient, $K_d$.

| 0.1 mM | Initial pH = 5.8 | | | |
|---|---|---|---|---|
| | Soil 1 | Soil 2 | Soil 3 | Soil 4 |
| Sample mass (g) | 49.9968 | 49.9816 | 50.0403 | 49.9698 |
| 0 h | 0.106 | 0.105 | 0.108 | 0.107 |
| 1 h | 0.0805 | 0.0654 | 0.0679 | 0.0422 |
| 2 h | 0.0777 | 0.0645 | 0.0668 | 0.0414 |
| 4 h | 0.0667 | 0.0594 | 0.0708 | 0.0404 |
| 8 h | 0.0683 | 0.0608 | 0.0699 | 0.0407 |

| 0.1 mM | Initial pH = 5.8 | | | |
|---|---|---|---|---|
| | Soil 1 | Soil 2 | Soil 3 | Soil 4 |
| Sample mass (g) | 49.9968 | 49.9816 | 50.0403 | 49.9698 |
| 24 h | 0.0618 | 0.0457 | 0.0491 | 0.0280 |
| 48 h | 0.0505 | 0.0363 | 0.0338 | 0.0277 |
| 96 h | 0.0418 | 0.0282 | 0.0213 | 0.0174 |
| 168 h | 0.0340 | 0.0236 | 0.0174 | 0.0126 |
| 336 h | 0.0292 | 0.0213 | 0.0135 | 0.00839 |
| 504 h | 0.0281 | 0.0208 | 0.0128 | 0.00871 |
| 672 h | 0.0243 | 0.0220 | 0.0104 | 0.00705 |
| $K_{d,\ adsorption}$ (l/kg) | 66.9 | 75.3 | 188 | 285 |
| 1.0 mM | Initial pH = 5.6 | | | |
| Sample mass (g) | 49.9897 | 50.0305 | 50.0279 | 50.0083 |
| 0 h | 1.17 | 1.13 | 1.16 | 1.13 |
| 1 h | 0.961 | 0.873 | 0.819 | 0.765 |
| 2 h | 0.944 | 0.859 | 0.800 | 0.752 |
| 4 h | 0.981 | 0.861 | 0.711 | 0.725 |
| 8 h | 0.878 | 0.843 | 0.709 | 0.723 |
| 24 h | 0.835 | 0.745 | 0.648 | 0.715 |
| 48 h | 0.788 | 0.665 | 0.509 | 0.650 |
| 96 h | 0.726 | 0.690 | 0.407 | 0.557 |
| 168 h | 0.710 | 0.612 | 0.350 | 0.470 |
| 336 h | 0.711 | 0.536 | 0.286 | 0.382 |
| 504 h | 0.671 | 0.452 | 0.276 | 0.370 |
| 672 h | 0.683 | 0.442 | 0.262 | 0.316 |
| $K_{d,\ adsorption}$ (l/kg) | 14.4 | 31.1 | 68.0 | 51.7 |
| 10 mM | Initial pH = 5.5 | | | |
| Sample mass (g) | 50.0059 | 49.9770 | 50.0090 | 50.0043 |
| 0 h | 11.7 | 11.5 | 11.6 | 11.6 |
| 1 h | 11.3 | 10.9 | 10.2 | 10.3 |
| 2 h | 11.2 | 10.7 | 10.0 | 10.5 |
| 4 h | 11.3 | 10.6 | 9.90 | 10.5 |
| 8 h | 11.2 | 10.7 | 9.68 | 10.3 |
| 24 h | 10.9 | 10.5 | 9.11 | 10.0 |
| 48 h | 10.9 | 10.3 | 8.79 | 9.73 |
| 96 h | 10.9 | 10.0 | 8.19 | 9.50 |
| 168 h | 10.8 | 10.0 | 8.10 | 8.32 |
| 336 h | 10.2 | 9.83 | 8.18 | 7.89 |
| 504 h | 9.92 | 9.06 | 8.15 | 8.33 |
| 672 h | 9.88 | 8.96 | 8.13 | 8.18 |
| $K_{d,\ adsorption}$ (l/kg) | 3.61 | 5.69 | 8.63 | 8.35 |

Table 4. Summary of the time-dependent Cs concentrations in solutions during the adsorption experiments. The initial concentrations were determined to be ~0.11, ~1.13, and ~11.6 mM, although they were prepared aiming $10^{-1}$, $10^{0}$, and $10^{1}$ mM.

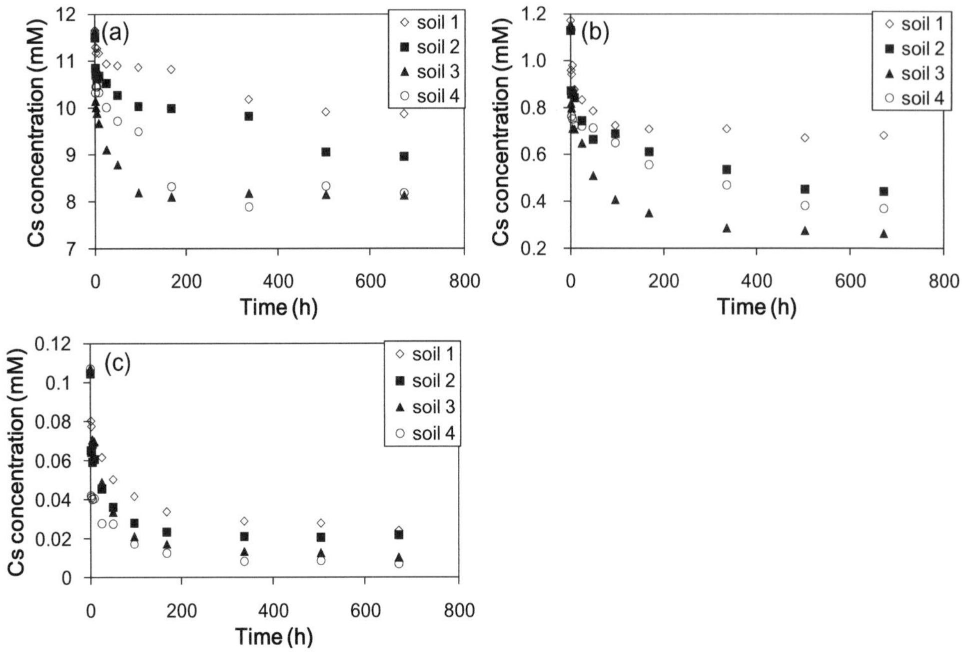


Fig. 5. Time-dependence of the Cs concentrations during the absorption experiments.

### 2.4.3 Characterization of Cs-adsorbed soils by electron microscopy

A rigorous characterization of the individual soil minerals was completed on soils 1, 3 and 4 interacted with the high Cs solution (~10 mM) using SEM. A SEM-EDX elemental map of the soil 1 sample revealed quartz, feldspar, and sheet-like aluminosilicate, which are consistent with the XRD results (Fig. 3). Cesium appears to localize only on the sheet-structured aluminosilicate. Figure 7 is the magnified image of the area indicated by the white squre in Fig. 6. The EDX point analysis indicated by the white circle revealed that the association of apatite and chloritized-biotite that contains a small amount of Cs.

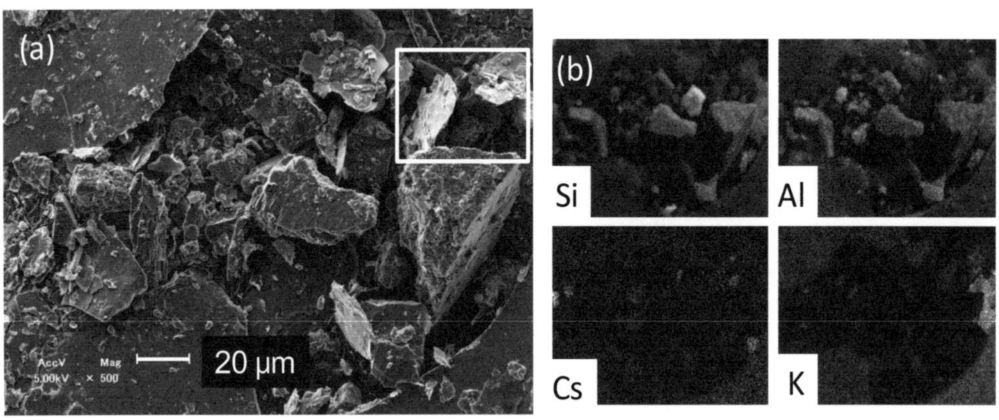


Fig. 6. Soil 1 sample. (a) Secondary electron image in SEM. (b) EDX elemental maps of the whole view in (a).

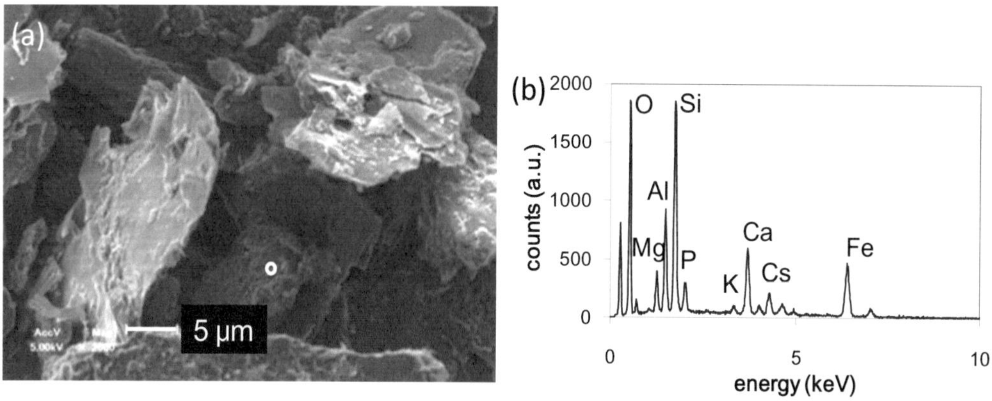


Fig. 7. The magnified image of the area indicated by the white square in Fig. 6. (a) SEM image. (b) The EDX spectrum obtained from the position indicated by the white circle.

On the other hand, the SEM-EDX elemental map and the EDX analysis of soil 3 showed the association of Cs with aggregates of sheet-structured aluminosilicate with illite composition (Fig. 8). Most of these illite particles were in the size <~20 μm, which are smaller than the chloritized biotite particles associated with Cs found in soil 1.

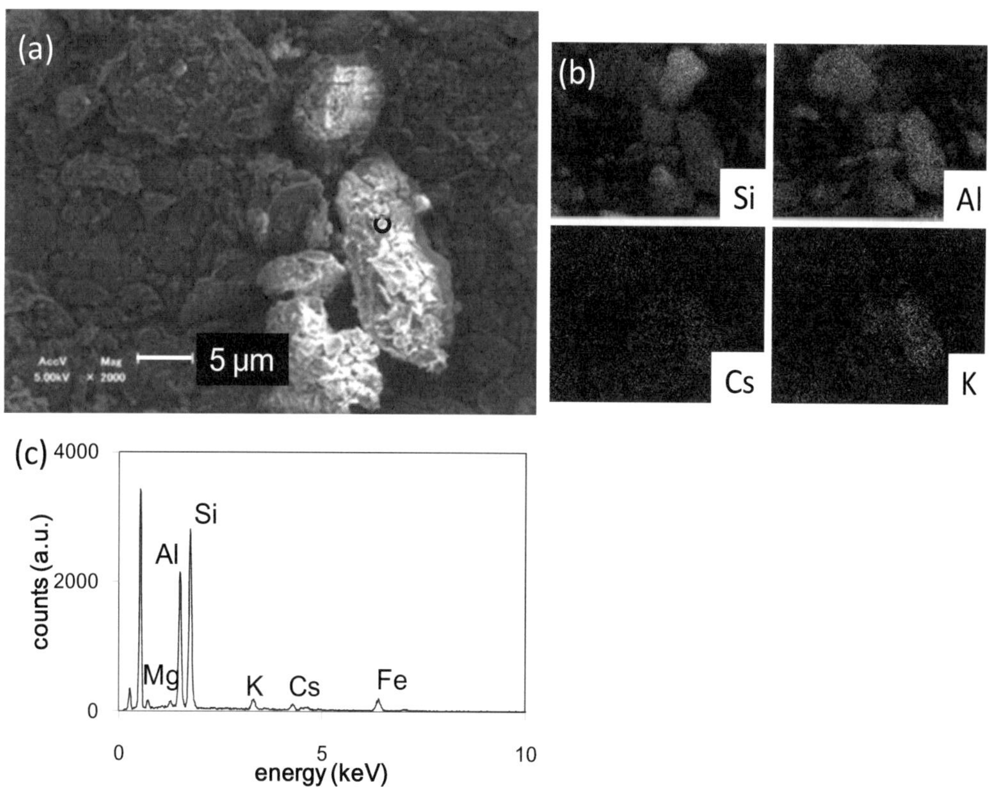


Fig. 8. SEM analysis of soil 3. (a) A secondary electron image. (b) EDX elemental maps of the view in (a). (c) The EDX spectrum of the point analysis indicated by the black open circle in (a).

In soil 4, sheet-structured aluminosilicate at the size about several hundred micron was abundant (Fig. 9). The composition of the sheet-silicates was mostly chlorite or chloritized biotite similar to the one observed in soil 1. Figure 10 is the close-up image of a biotite particle associated with the elemental maps and the EDX point analyses. The elemental map clearly shows Cs concentration at the edge of the biotite paritcles, which is the evidence of favourable Cs adsorption at the frayed-edge site in addition to the homogeneous distribution in/on the particle. The EDX line analyses conducted on a particle clearly indicate that Cs incorporation is associated with K depletion at the frayed-edge (Fig. 11). The direct evidence of Cs concentraion at the frayed-edge was also reported for biotite (McKinley et al., 2004; Wang et al., 2010) and micas (Liu et al., 2003). It is also worth noting the presence of cleavage plane and the distribution of numerous smaller particles attaching on the surface of the platy particle (Fig. 11). These small particles are associated with moderately high concentration of Cs as shown in STEM-EDX maps (Fig. 12), which account for the Cs concentration in the SEM point analyses or the SEM-EDX maps.

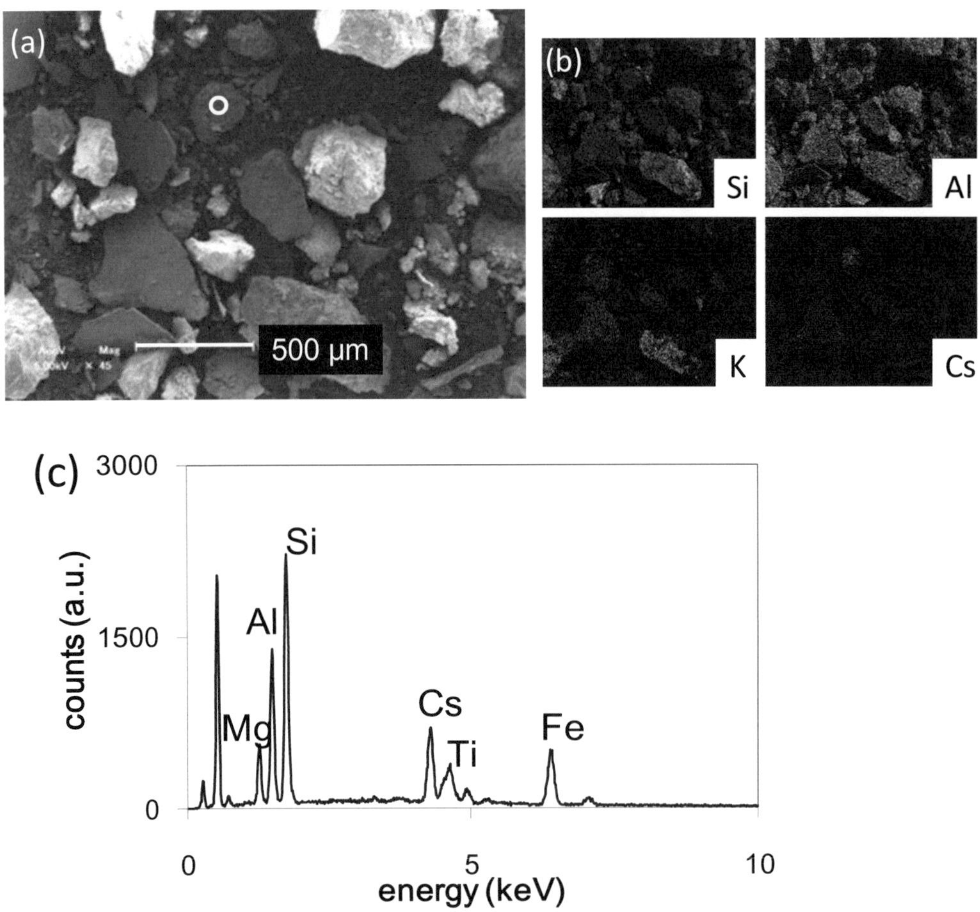


Fig. 9. Soil 4 sample. (a) A secondary electron image. (b) SEM-EDX elemental maps of the view of (a). (c) The EDX spectrum obtained from the position indicated by the white open circle in (a).

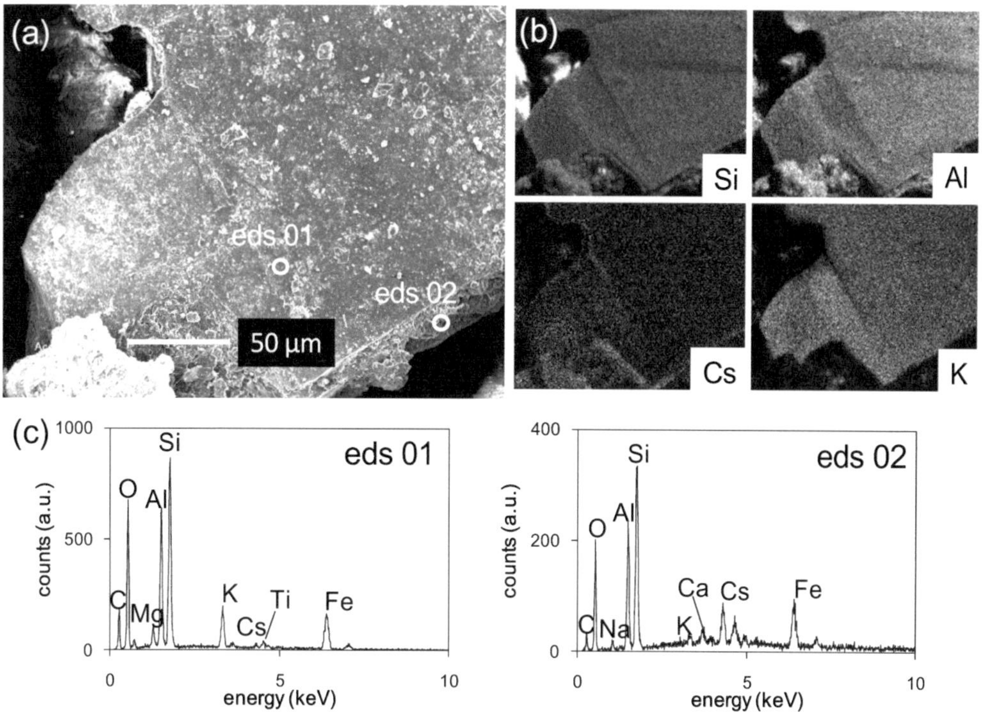


Fig. 10. Soil 4 sample. (a) Secondary electron image of a biotite particle. (b) SEM-EDX elemental maps maps. (c) The EDX spectrums at the points eds 01 and eds 02 indicated in (a).

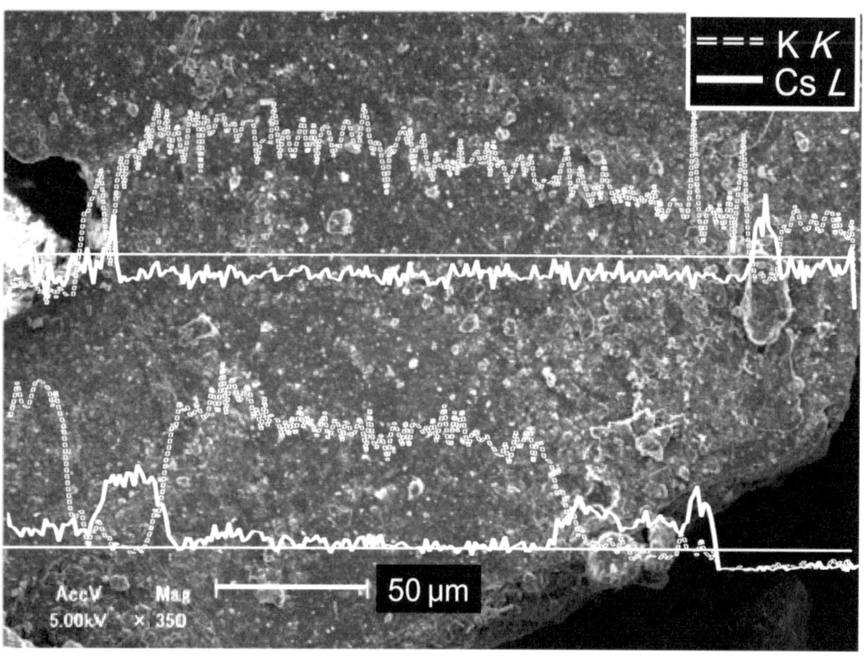


Fig. 11. EDX line scan ananlyses of the two traverses across a chloritized biotite particle.

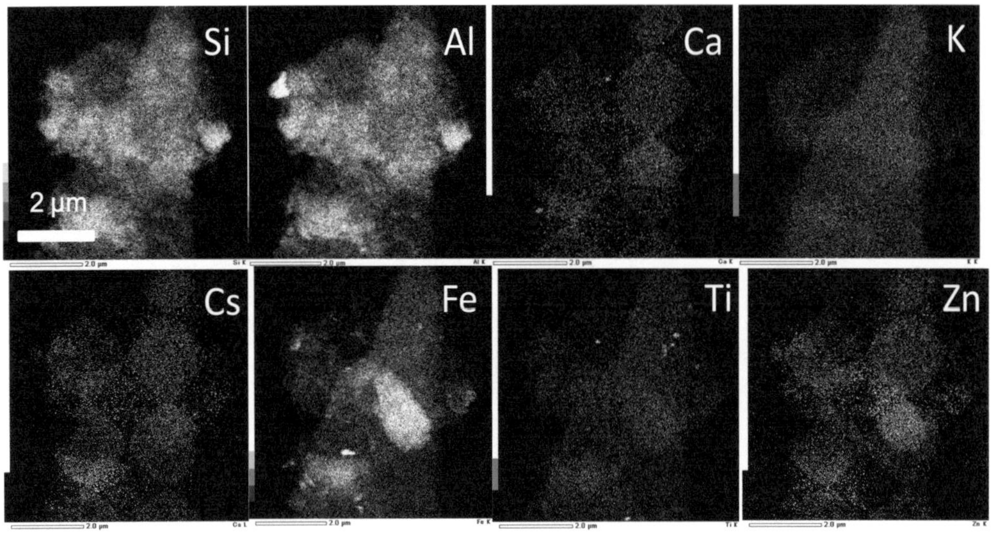


Fig. 12. STEM-EDX elemental maps of a chloritized biotite particle.

Overall, chlorite and chloritized biotite were major Cs adsorptive phase in a soils 1 and 4. Semiquantification of these minerals indicated a wide range of K concentration in chloritized biotite; however, there is a trend that Cs concentration increases as the K concentration decreases (Fig. 13). That is, chlorite is capable of adsorbing greater amount of Cs than biotite. (Fig. 13).

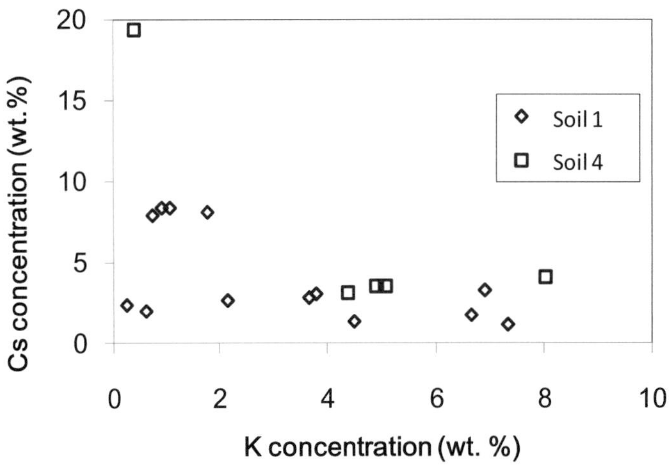


Fig. 13. The Cs concentration in biotite/chloritized biotite/chlorite particles plotted as a function of K concentration based on the semi-quantitative SEM-EDX point analysis.

### 2.4.4 Cesium desorption in extractants

After the adsorption experiments, the selected Cs-sorbed samples (soils 1 and 3 preliminarily contacted with 10 mM CsCl solution) were subsequently used for the desorption experiments. The time-dependent Cs concentrations in solutions were summarized in Table 5 accompanied with pHs and $K_{d,desorption}$s. The $K_{d,desorption}$ was calculated by the following equation:

$$K_{d,desorption} = \frac{(C_i - C_f) - C_f'}{C_f'} \times \frac{W_l'}{W_s'}$$

$C_f'$ : the final Cs concentration (mol / l)
$W_l'$ : the liquid solution volume (l)
$W_s'$ : the soil mass (kg)

Soil 1

| Extractant | 0.1 M $MgCl_2$ | 0.1 M citric acid | 0.1 M acetic acid | 0.1 M KCl | 0.1 M $NH_4Cl$ | MilliQ |
|---|---|---|---|---|---|---|
| Initial pH | 9.1 | 2.0 | 2.8 | 5.4 | 5.1 | 6.3 |
| Sample mass (g) | 5.0032 | 5.0204 | 5.0182 | 5.0085 | 5.0024 | 5.0067 |
| 0 h | < 0.0001 | < 0.0001 | < 0.0001 | < 0.0001 | < 0.0001 | < 0.0001 |
| 1 h | 0.190 | 0.209 | 0.186 | 0.276 | 0.248 | 0.025 |
| 2 h | 0.206 | 0.231 | 0.210 | 0.298 | 0.262 | 0.028 |
| 4 h | 0.225 | 0.250 | 0.222 | 0.326 | 0.279 | 0.031 |
| 8 h | 0.223 | 0.270 | 0.237 | 0.330 | 0.282 | 0.043 |
| 24 h | 0.239 | 0.275 | 0.256 | 0.370 | 0.291 | 0.045 |
| 48 h | 0.250 | 0.296 | 0.253 | 0.360 | 0.279 | 0.046 |
| 96 h | 0.250 | 0.282 | 0.254 | 0.354 | 0.287 | 0.047 |
| 168 h | 0.253 | 0.304 | 0.262 | 0.339 | 0.281 | 0.049 |
| Final pH | 6.6 | 2.1 | 3.0 | 4.9 | 4.9 | 5.7 |
| $K_{d,desorption}$ (l/kg) | 242.5 | 194.4 | 231.7 | 170.6 | 213.8 | 1410 |

Soil 3

| | | | | | | |
|---|---|---|---|---|---|---|
| Initial pH | 9.1 | 2.0 | 2.8 | 5.4 | 5.1 | 6.3 |
| Sample mass (g) | 5.0011 | 5.0087 | 5.0176 | 5.0107 | 5.0065 | 5.0056 |
| 0 h | < 0.0001 | < 0.0001 | < 0.0001 | < 0.0001 | < 0.0001 | < 0.0001 |
| 1 h | 0.994 | 0.691 | 0.635 | 1.339 | 1.233 | 0.210 |
| 2 h | 1.020 | 0.805 | 0.687 | 1.476 | 1.300 | 0.212 |
| 4 h | 1.198 | 0.872 | 0.751 | 1.593 | 1.381 | 0.235 |
| 8 h | 1.249 | 0.968 | 0.870 | 1.683 | 1.428 | 0.251 |
| 24 h | 1.317 | 1.101 | 0.855 | 1.744 | 1.499 | 0.283 |
| 48 h | 1.339 | 1.100 | 1.012 | 1.729 | 1.499 | 0.289 |
| 96 h | 1.340 | 1.088 | 1.058 | 1.730 | 1.487 | 0.300 |
| 168 h | 1.319 | 1.177 | 1.132 | 1.756 | 1.498 | 0.313 |
| Final pH | 5.0 | 2.1 | 3.1 | 4.2 | 4.2 | 4.8 |
| $K_{d,desorption}$ (l/kg) | 66.39 | 79.11 | 83.66 | 39.81 | 53.59 | 407.3 |

Table 5. Summary of the time-dependent Cs concentration during the desorption experiments on 10 mM Cs-sorbed soil 1 and soil 3, accompanied by the solution pHs and $K_{d,desorption}$.

$K_{d,desorption}$ in soil 1 is greater than that in soil 3, most likely due to the greater amount of loaded Cs on the soils 3 in the adsorption experiment, which is consistent with a previous work (Wang et al., 2010). The Cs desorption efficiency of extractants is in the order of KCl > citric acid > $NH_4Cl$ > acetic acid > $MgCl_2$ >> DW in soil 1, whereas KCl > $NH_4Cl$ > $MgCl_2$ > citric acid > acetic acid >> DW in the highly loaded soil 3. This difference may indicate the change in efficiency depending on the amount of loaded Cs. The order found in the less Cs-loaded soil 1 is similar to the one previously reported; sea water > groundwater > sodium acetate ~ $MgCl_2$ > DW (Wang et al. 2010). Another previous study has reported that citric acid is the most effective extractant among low molecular weight organic acids owing to its three carboxyl ligands (Chiang et al., 2011). The authors suggested that protonation first occurred and the organic ligands subsequently attack OH and $H_2O$ group. On the other hand, our results clearly showed the highest efficiency in the use of KCl solution, implying the importance of cation-exchange and diffusion mechanism through the interlayer of clay minerals. Figures 14a and b are the plotted data of the Cs concentrations as a function of time, indicating that apparent equilibrium was achieved only after 24 h, which is much faster than the case in the adsorption experiment. Chiang et al. (2011) proposed that inter- and intra-layer diffusion may be dominant during the first 24 hours.

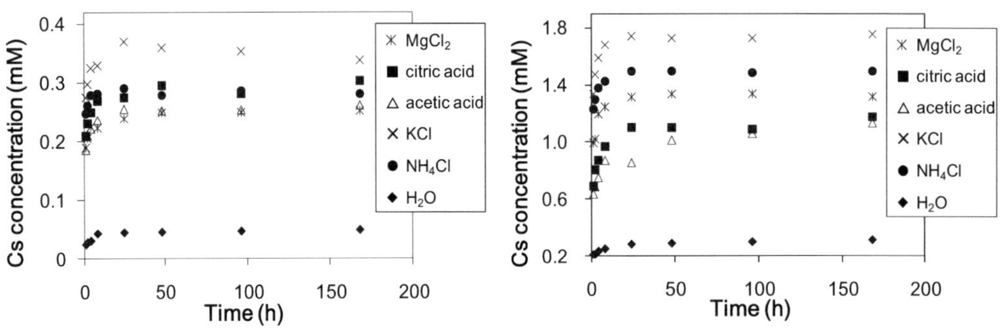


Fig. 14. Time-dependent Cs concentration during the desorption experiments.

| | Soil 1 | Soil 3 |
|---|---|---|
| $MgCl_2$ | 0.0093 | 0.0157 |
| citric acid | 0.0105 | 0.0125 |
| acetic acid | 0.0091 | 0.0083 |
| KCl | 0.0131 | 0.0178 |
| $NH_4Cl$ | 0.0082 | 0.0106 |
| $H_2O$ | 0.0016 | 0.0019 |

Table 6. The apparent rate constant of Cs desorption calculated based on the data from soil 1 and soil 3. (% $sec^{-1}$).

For the quantitative comparision of efficiency of the extractants, the apparent desorption rate was calculated by drawing a regression line on the data at 1, 2, and 4 h, of which the unit is written as mM $h^{-1}$. This apparent desorption rate was further converted to the dimension of second and then calibrated by the total amount of Cs on the soil in order to obtain the proportion of the amount of desorbed Cs to that of adsorbed Cs in %, which has a unit of % $sec^{-1}$. The results obtained by the conversion clearly revealed that the apparent Cs desorption rate is the fastest in the case of the use of KCl. Thus, it is concluded that KCl is the most effective extractant for the Cs decontamination of the Fukushima soils. However, the counter anion $Cl^-$ may cause a chloride damage. Hence, the K-bearing chemical compounds with minimum toxicity should be explored in the future research.

## 3. Conclusions

In this chapter, we have reviewed major events related to the accident occurred at the FDNPP. A variety of fissiogenic radionuclides were released to the environment, among which $^{137}$Cs is the most critical contaminant in the surface soils in the vicinity of the FDNPP. Thus, it is highly demanded to establish the efficient protocol of Cs decontamination. The present study demonstrated Cs adsorption experiments using four different types of soils collected from Fukushima. The rigorous characterization of the Cs-adsorbed soils showed that Cs is associated with both high- and low-affinity sites of illite, muscovite, biotite, chloritized biotite, and chlorite. Cesium desorption experiments were also conducted using six extractants: DW, 0.1 mol/l KCl, $NH_4Cl$, $MgCl_2$, acetic acid, and citric acid. The results clearly revealed the highest efficiency of Cs removal when using KCl solution, suggesting that the K-based chemical compound is a key extractant for site remediation in Fukushima.

## 4. Acknowledgments

We are grateful to the staff of the Center of Advanced Instrumental Analysis and HVEM of Kyushu University for the technical support in TEM, SEM and ICP-MS analyses. This work was financially supported by the ESPEC Foundation for Global Environment Research and Technology (Charitable Trust) (ESPEC Prize for the Encouragement of Environmental Studies) and partly supported by Basic Research Funds from the Radioactive Waste Management Funding and Research Center.

## 5. References

Bellenger, J. P. & Staunton, S. (2008) Adsorption and desorption of $^{85}$Sr and $^{137}$Cs on reference minerals, with and without inorganic and organic surface coatings. Journal of Environmental Radioactivity, 99, 831-840.

Bouzidi, A.; Souahi, F.; Hanini, S. (2010) Sorption behavior of cesium on Ain Oussera soil under different physicochemical conditions. Journal of Hazardous Materials, 184, 640-646.

Chiang, P. N.; Wang, M. K.; Huang, P. M. & Wang, J. J. (2011) Effects of low molecular weight organic acids on $^{137}$Cs release from contaminated soils. Applied Radiation and Isotopes, 69, 844-851.

Chino, M.; Ishikawa, H. & Yamazawa et al. (1986) Application of the SPEEDI system to the Chernobyl reactor accident. JAERI-M 86-142.

Chino, M.; Nakayama, H.; Nagai, H.; Terada, H.; Karata, G. & Yamazawa, H. (2011) Preliminary estimation of release amount of $^{131}$I and $^{137}$Cs accidentally discharged from the Fukushima Daiichi Nuclear Power Plant into the atmosphere. Journal of Nuclear Science Technology, 48, 1129-1134.

Comans, R. N. J. & Hockley, D. E. (1992) Kinetics of cesium sorption on illite. Geochimica et Cosmochimica Acta, 56, 1157-1164.

Hassan, N. M. (2005) Adsorption of cesium from spent nuclear fuel basin water. Journal of Radioanalytical and Nuclear Chemistry, 266, 57-59.

Hsu, C. N. & Chang, K. P. (1994) Sorption and desorption behavior of cesium on soil components. Applied Radiation and Isotopes, 45, 433-437.

Khan, S. A.; Rehman, R. U. & Kahn, M. A. (1994) Sorption of cesium on bentonite. Waste Management, 14, 629-642.

Liu, C.; Zachara, J. M.; Smith, S. C.; McKinley, J. P. & Ainsworth, C. C. (2003) Desorption kinetics of radiocesium from subsurface sediments at Hanford Site, USA. Geochimica et Cosmochimica Acta, 67, 2893-2912.

McKinley, J. P.; Zachara, J. M.; Heald, S. M.; Dohnalkova, A.; Newville, M. G. & Sutton S. R. (2004) Microscale distribution of cesium sorbed to biotite and muscovite. Environmental Science & Technology, 38, 1017-1023.

MEXT (2001) http:/www.mext.go.jp/a_menu/saigaijohou/syousai/1305747.htm. Database of Ministry of Education, Culture, Sports, Science and Technology, Japan,

Missana, T.; Garcia-Gutierrez, M. & Alonso, U. (2004) Kinetics and irreversibility of cesium and uranium sorption onto bentonite colloids in a deep granitic environment. Applied Clay Science, 26, 137-150.

Mon, J.; Deng, Y.;l Flury, M. & Harsh, J. B. (2005) Cesium incorporation and diffusion in cancrinite, sodalite, zeolite, and allophane. Microporous and Mesoporous Materials, 86, 277-286.

Naito, M. (2011) Damage of reactor buildings at Fukushima Daiichi Nuclear Power Plants – Why hydrogen explosion occurred. Journal of Atomic Energy Society of Japan, 53, 473-478 (in Japanese).

Narabayashi, T. & Sugiyama, K. (2011) Fukishim 1st NPPs Accidents and Disaster Caused by the Pacific Coast Tsunami of Tohoku Earthquake; Lessons from evaluation of the Fukushima 1st NPPs accidents. Journal of Atomic Energy Society of Japan, 53, 387-400 (in Japanese).

NSCJ (2011) http://www.nsc.go.jp/info/20110412. Database of Nuclear Safety Commission, Japan.

Rajec, P.; Sucha, V.; Eberl, D. D.; Srodon, J. & Elsass, F. (1999) Effect of illite particle shape on cesium sorption, Clays and Clay Minerals, 47, 755-760.

Rajec, P. & Domianova, K. (2008) Cesium exchange reaction on natural and modified clinoptilolite zeolites. Journal of Radioanalytical and Nuclear Chemistry, 275, 503-508.

Seaman, J. C.; Meehan, T. & Bertsch, P. M. (2001) Immobilization of Cesium-137 and uranium in contaminated sediments using soil amendments. Journal of Environmental Quality, 30, 1206-1213.

Steefel, C. I.; Carroll, S.; Zhao, P. & Roberts, S. (2003) Cesium migration in Hanford sediment: a multisite cation exchange model based on laboratory transport experiments. 67, 219-246.

TEPCO (2011) http://www.tepco.co.jp/nu/fukushima-np/f1/index-j.html. Database of the Tokyo Electric Power Company.

Wang, T. H.; Li, M. H.; Wei, Y. Y. & Teng, S. P. (2010) Desorption of cesium from granite under various aqueous conditions. Applied Radiation and Isotopes, 68, 2140-2146.

Yildiz, B.; Erten, H. N. & Kis, M. (2011) The sorption behaviour of $Cs^+$ ion on clay minerals and zeolite in radioactive waste management: sorption kinetics and thermodynamics. Journal of Radioanalytical and Nuclear Chemistry, 288, 475-483.

Zachara, J. M.; Smith, S. C.; Liu, C.; McKinley, J. P.; Serne, R. J. & Gassman, P. L. (2002) Sorption of Cs+ to micaceous subsurface sediments from the Hanford site, USA. Geochimica et Cosmochimica Acta, 66, 193-211.

# Behaviors of Nuclear Fuel Cladding During RIA

Sun-Ki Kim
*Korean Atomic Energy Research Institute*
*Republic of Korea*

## 1. Introduction

A Reactivity-initiated accident (RIA) is a nuclear reactor accident that involves an unwanted increase in reactor power. The abrupt power increase can lead to damage the reactor core as well as fuel cladding, and in severe cases, even lead to disruption of reactor.

During a steady-state operation of light water reactors, the mechanical behavior of the zirconium-based fuel cladding degrades due to a combination of oxidation, hydriding, and radiation damage. In an effort to increase the operating efficiency through the use of longer fuel cycles, and to reduce the volume of waste associated with the core reloads, utilities have a strong incentive to increase the average discharge burn-up of the fuel assemblies. Further increases in the operating efficiency of power reactors can also be achieved by increasing the coolant outlet temperature. However, both of these changes in a reactor operation enhance the cladding degradation, which may increase the likelihood of a cladding failure during design-basis accidents.

One such postulated design-basis accident scenario is the reactivity-initiated accident (RIA) in a pressurized water reactor (PWR) caused by the ejection of a control rod from the core, which would cause a rapid increase of the reactivity and the thermal energy in the fuel (Meyer et al., 1986). The increase in fuel temperature resulting from an RIA induces a rapid fuel expansion, causing a severe pellet-cladding mechanical interaction (PCMI). This PCMI forces the cladding to experience a multiaxial tension such that the maximum principal strain is in the hoop (i.e., transverse) direction of the cladding tube. The survivability of a fuel cladding irradiated to a high burn-up under postulated RIA conditions is thus a response to a combination of the mechanics of a loading and the material degradation during a reactor operation.

While such data is available for the axial deformation behavior of cladding tubes, relatively little data has been reported in the open literature on the uniaxial tension behavior in the hoop direction of nuclear fuel cladding. In the 1990s, experimental programs were also initiated in Japan, France, and Russia to investigate the behavior of highly irradiated nuclear fuel under reactivity-initiated accident conditions.

Accordingly, it is essential to investigate the uniaxial tension behavior in the hoop direction of nuclear fuel cladding. In this chapter, some mechanical tests results for simulating the cladding behaviors during reactivity-initiated accident are introduced.

## 2. International test program: CABRI program

In a pressurized water reactor, the reactivity initiated accident (RIA) scenario of primary concern is the control rod ejection accident (Glasstone & Sesonske, 1991). At this early heat-up stage of the RIA, the clad tube material is still at a fairly low temperature (<650 K), and the fast straining imposed by the expanding fuel pellets may therefore cause a rapid and partially brittle mode of clad failure (Chung & Kassner, 1998). To investigate fuel behaviors during these RIA condition, CABRI REP-Na test program was initiated in early 1990s. The main purpose of the CABRI REP-Na test program was to study the validity of the RIA acceptance criteria on high-burnup 17×17 PWR fuel, with emphasis on the behavior of fuel during the early stage of the transient up to fuel failure (Schmitz & Papin, 1999). Recently, international test program in CABRI has being prepared in order to simulate the fuel behaviors in water loop at reactor pressure level by replacing the existing Na loop with water loop. Through a total of 15 technical advisory group meetings, the test matrix in the CABRI water loop program has being established.

In this chapter, main test results in CABRI REP-Na program are presented, and also the current status on the CABRI water loop program is briefly introduced.

### 2.1 CABRI REP Na loop test program

#### 2.1.1 Overview of CABRI REP Na loop test

First of all, a total of twelve tests have been carried out in the program. Eight of the tests were performed on $UO_2$ fuel and four tests on MOX fuel, pre-irradiated to burnups ranging from 28 to 65 MWd/kgU. In addition to the CABRI REP-Na program, two tests were performed in November 2002 on PWR fuel rods with burnups around 75 MWd//kgU. These test rods, denoted CIPO-1 and CIPO-2, had advanced claddings such as ZIRLO and M5 cladding, respectively. The CABRI test reactor is a pool-type light water reactor, designed with a central flux area that can accommodate the insertion of a test device.

### 2.2 Main outcomes of CABRI REP Na loop test

Fuel cladding failure occurred in four tests, whereas no failure occurred in the remaining 8 tests. Among the four tests with fuel failure, there had $UO_2$ fuel and one MOX fuel. All failures were with Zircaloy-4 cladding, whereas no failure occurred in the three tests with M5 or ZIRLO cladding. The three $UO_2$ failures occurred at enthalpy below 80 cal/g and on fuel that had a burnup of about 60 MWd/kgU and significant oxide thickness from 80 to 130 μm. However, several other CABRI tests were run at comparable burnup and oxide thickness range, without resulting in fuel failure. Oxide spalling appears to be distinctive element that separates the failed and non-failed fuel in CABRI $UO_2$ tests, in that all three fuel rods that failed had spalling, while the rods that did not fail had uniform non-spalled oxide. Only one test for MOX fuel among four tests resulted in fuel failure. The failure occurred on a fuel rod that had a burnup of 55 MWd/kgU and moderate corrosion level of 50 μm oxide thickness. This is the only CABRI failure occurring on a non-spalled cladding. However, this failure occurred at 113cal/g, which is a rather high enthalpy level, where a failure may not be surprising, considering that this level in near the upper envelope of all CABRI data. Failures were not associated to a particular pulse width. The three $UO_2$ fuel rod failures occurred at all three pulse widths that have been used in CABRI test program, i.e. 9, 30, and 75 ms. The main outcomes from CABRI Na loop test program are listed in Table 1.

### 2.3 Current status of CABRI water loop test program

CABRI international program (CIP) in water loop test has been managed by IRSN in France with collaboration of EDF, CEA and the participating countries with support by OECD/NEA. This program focuses on the high burnup fuel and cladding behaviors in PWR condition during RIA. The test conditions are 155 bar and 280°C. Five test series were

| Test and date | Rod and Burn-up | Pulse width (ms) | Energy Deposition cal/g | Corrosion μm | Results and observations |
|---|---|---|---|---|---|
| Na-1 (11/93) | GRA 5 64 MWd/kg | 9.5 | 110 (at 0.4s) | 80 spalling | Brittle failure at $H_F$ = 30 cal/g Fuel dispersal (6g) |
| Na-2 (6/94) | BR3 33 MWd/kg | 9.1 | 211 (at 0.4s) | 4 | No failure, $H_{MAX}$ = 199 cal/g Max. strain : 3.5%, FGR : 5.5% |
| Na-3 (10/94) | GRA 5 53 MWd/kg | 9.5 | 120 (at 0.4s) | 40 | No failure, $H_{MAX}$ = 124 cal/g Max. strain : 2%, FGR : 13.7% |
| Na-4 (7/95) | GRA 5 62 MWd/kg | 75 | 95 (at 1.2s) | 80 no spalling | No failure, $H_{MAX}$ = 85 cal/g Max. strain : 0.4%, FGR : 8.3% |
| Na-5 (5/95) | GRA 5 64 MWd/kg | 9.5 | 105 (at 0.4s) | 20 | No failure, $H_{MAX}$ = 108 cal/g Max. strain : 1%, FGR : 15.1% |
| Na-6 (03/96) | MOX 47 MWd/kg | 35 | 125 (at 0.66s) | 35 | No failure, $H_{MAX}$ = 133 cal/g Max. strain : 3.2%, FGR : 21.6% |
| Na-7 (1/97) | MOX 55 MWd/kg | 40 | 165 (at 1.2s) | 50 | Failure at $H_F$ = 113 cal/g Strong flow ejection |
| Na-8 (07/97) | GRA 5 60 MWd/kg | 75 | 106 (at 0.4s) | 130 lim. spalling | Failure at $H_F \leq$ 82 cal/g $H_{MAX}$ = 98 cal/g, No fuel dispersal |
| Na-9 (04/97) | MOX 28 MWd/kg | 34 | 197 at 0.5s 241 at 1.2s | ⟨20 | No failure, $H_{MAX}$ = 197 cal/g Max. strain : 7.4%, FGR : ~34% |
| Na-10 (07/98) | GRA 5 62 MWd/kg | 31 | 107 at 1.2s | 80 no spalling | Failure at $H_F$ = 81 cal/g $H_{MAX}$ = 98 cal/g, No fuel dispersal |
| Na-11 | M5 63 MWd/kg | 31 | 104 | 15 | No failure, $H_{MAX}$ = 93 cal/g Max. strain : ~0.5% |
| Na-12 | MOX 65 MWd/kg | 62 | 106 | 80 no spalling | No failure, $H_{MAX}$ = 103 cal/g |
| CIP0-1 (11/02) | ZIRLO 75 MWd/kg | 32 | 98 | 80 no spalling | No failure, $H_{MAX}$ = 90 cal/g |
| CIP0-2 (11/02) | M5 77 MWd/kg | 28 | 89 | 20 | No failure, $H_{MAX}$ = 81 cal/g |

Table 1. Main outcomes from CABRI REP Na test program

established including qualification test, CIP Q for various fuel pellet, cladding, corrosion level and burnup. The test matrix CABRI international program (CIP) in water loop are listed in Table 2. The test sequence is not fixed, therefore it may be changed.

| Test Series | Test | Test Rod | Objectives |
|---|---|---|---|
| CIPQ | | REP Na6 rod, span 3, 3cycles MOX, Zr4 | CWL qualification + phenomenology |
| CIP1 | *CIP1-2 (option)* | CIP0-2 sister rod, $UO_2$ 77GWd/t, M5 | Phenomenology, Code validation |
| CIP2 | CIP2-1 | $UO_2$ 85 GWd/t, M5 = reference or ENUSA rod , 80GWd/t, Zr-Nb | Product qualification, code validation |
| CIP4 | CIP4-1<br><br>CIP4-2 | MOX-E M5 EDF rod (55GWd/t)<br>MOX-SBR-Beznau rod (54Gwd/t, Zr4) | Product qualification<br>Product qualification, physical understanding |
| CIP5 | CIP5-1 | VVER Lovisa, 50GWd/t, Zr-1%Nb | Product qualification |
| CIP3 | CIP3-1 | CIP0-1 sister rod, span 5 | Code validation, safety criteria, post-failure events |
| | CIP3-2 | REP Na 7 sister rod | Code validation, safety criteria, post-failure events |
| | CIP3-3 | Vandellos CIP0-1 sister rod, span3, $UO_2$, 75 GWd/t | Code validation, safety criteria, phenomenology |
| | CIP3-4 | EDF rod, M5 63-68 GWd/t | Product qualification |
| | CIP3-5 | ENUSA rod, 67 GWd/t Zr-Nb | Product qualification, physical understanding |

Table 2. Test matrix CABRI international program

## 3. Evaluation of mechanical properties of nuclear cladding

Except for international test program in pulse reactors, some mechanical tests such as ring tensile test, high-speed burst test, and expansion due-to compression test were applied in order to investigate the stress and strain in pellet-cladding mechanical interaction during RIA. In this chapter, the results on the ring tensile test and the high-speed burst test are introduced.

### 3.1 Evaluation of mechanical properties by Ring tensile test

To obtain the mechanical strengths such as the 0.2% offset yield strength and ultimate tensile strength were evaluated, and the uniform elongation and total elongation were also evaluated for the ductility. The hoop stress-strain curves at various temperatures are shown in Fig. 1.

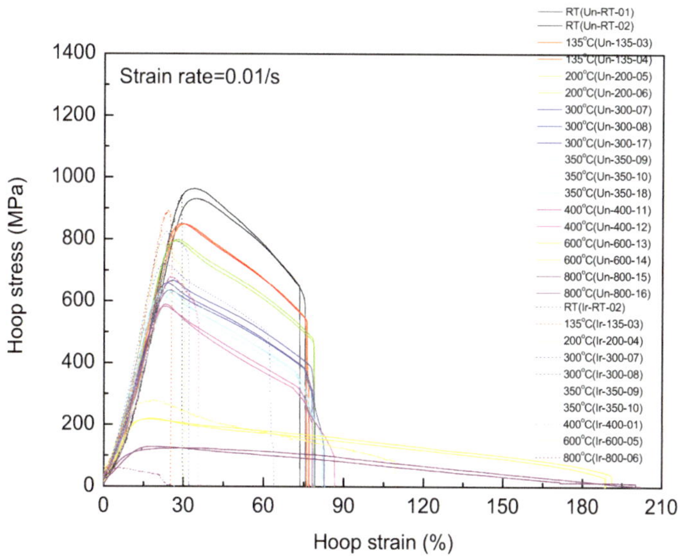


Fig. 1. Hoop stress-strain curves at various test temperatures

The evaluation results of the yield strength and the ultimate tensile strength are shown in Fig. 2 and 3.

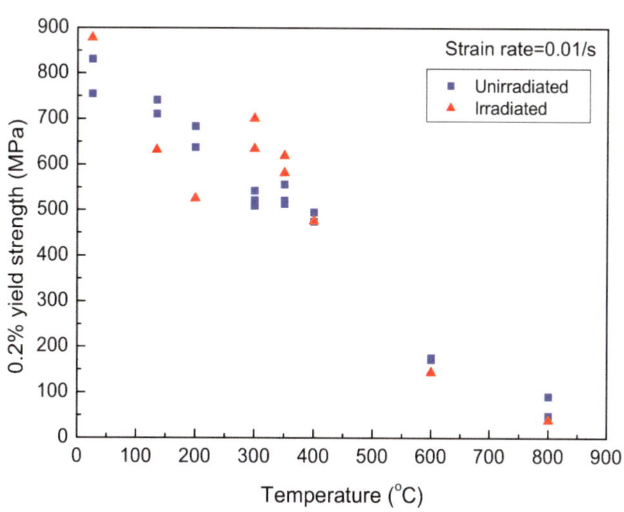


Fig. 2. Yield strength of the un-irradiated and high burn-up fuel cladding

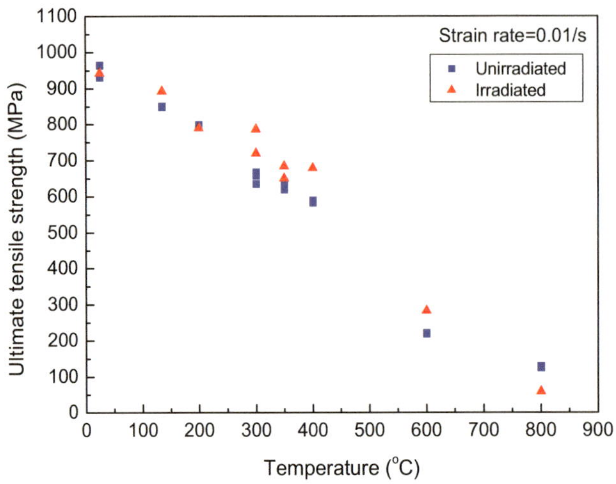


Fig. 3. Ultimate tensile strength of the un-irradiated and high burn-up fuel cladding

The results show that the 0.2% offset yield strength and the ultimate tensile strength abruptly decrease with an increasing temperature. The ultimate tensile strength was evaluated to be 942.70 MPa at RT, 678.83 MPa at 400°C, but, it is abruptly diminished to 282.64 MPa at 600°C, which is achievable in the RIA condition. Especially, it decreases to 58.30 MPa at 800°C, an extreme condition, which corresponds to 6% of the ultimate tensile strength at room temperature. This means that the mechanical strength of the high burn-up Zircaloy-4 nuclear fuel cladding sharply decreases in the RIA-relevant temperature ranges. The evaluation results of the uniform elongation and total elongation are shown in Fig. 4 and 5.

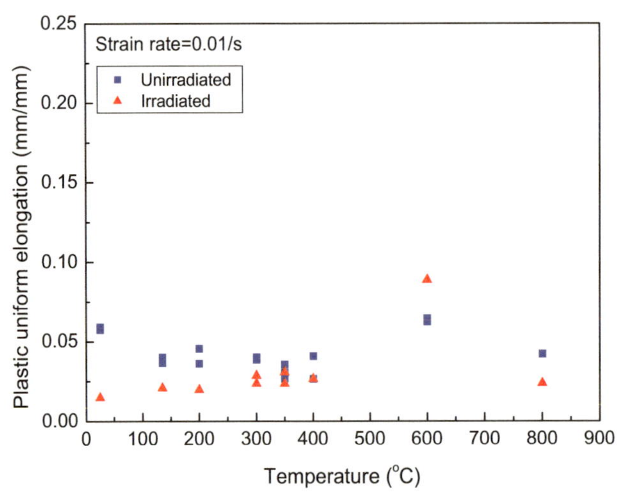


Fig. 4. Uniform elongation of the un-irradiated and high burn-up fuel cladding

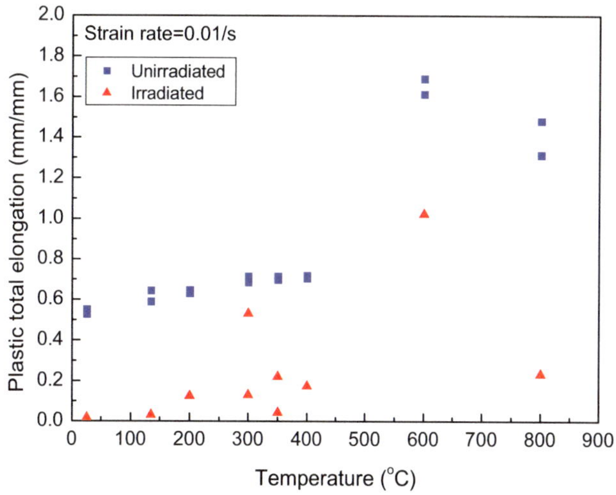


Fig. 5. Total elongation of the un-irradiated and high burn-up fuel cladding

The results show that both the increases in uniform elongation and total elongation with an increasing temperature. Especially, they abruptly increase at 600°C, but become lower beyond this temperature. This peculiar behavior was also observed in the PROMETRA program (Averty et al., 2003) which is a mechanical property relevant test program in conjunction with the CABRI program simulating a RIA. The results of hoop directional mechanical strength and ductility of irradiated Zircaloy-4 with comparison with PROMETRA database are shown in Fig. 6 and Fig. 7, respectively. As shown in the figure, the behavior of this study is consistent with the PROMETRA data. It is believed that this behavior is caused by the elongation minimum phenomenon by the dynamic strain aging of the Zr-base cladding material beyond 600°C.

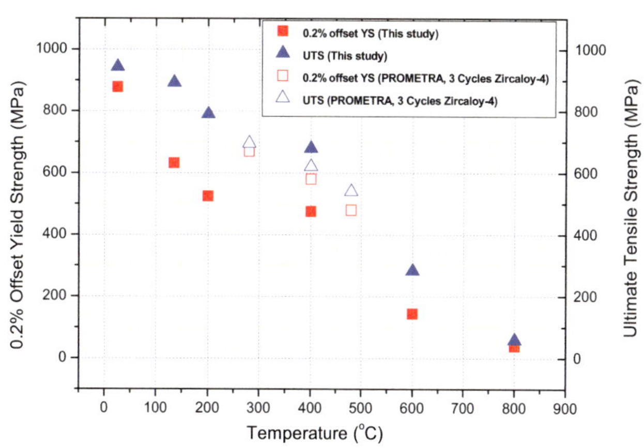


Fig. 6. Comparison of mechanical strength with the PROMETRA database

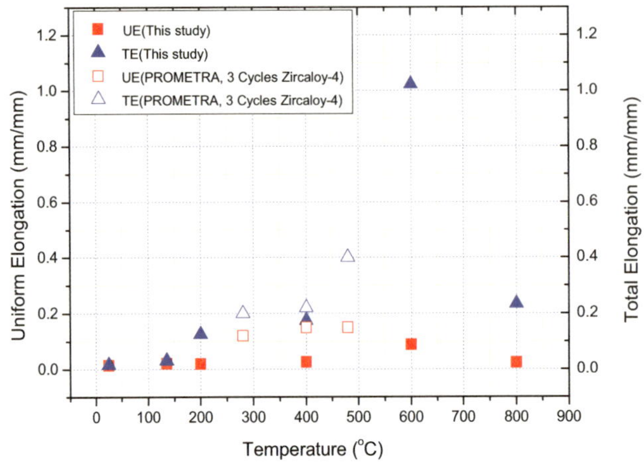


Fig. 7. Comparison of mechanical ductility with the PROMETRA database

The stereoscope photographs of the specimens after the ring tensile test are shown in Fig. 8. Fig. 8 represents the fracture patterns of the non-irradiated and high burn-up Zircaloy-4 cladding specimens. As shown in the figure, four fracture patterns such as 45° shear type fracture, cup & cone type fracture, cup & cup type fracture and chisel edge type fracture were observed in the non-irradiated cladding.

The fracture type was found to depend strongly on the deformation temperature. At room temperature non-irradiated Zircaloy-4 cladding tends to be fractured by 45° shear type fracture or Cup & cone type fracture. Cup & cone type fracture was dominant at 135°C and 200°C. Both cup & cone type fracture and cup & cup type fracture was observed at 300°C. Only cup & cup type fracture was observed at 350°C and 400°C. At 600°C chisel edge type fracture was observed unlike at lower temperatures with few fracture surface area. At 800°C the fracture type is unclear. These fracture types are summarized in Table 3.

| Temperature | Fracture type | |
|---|---|---|
| | ***Non-irradiated cladding*** | ***High burn-up cladding*** |
| 25°C | *45 ° shear type fracture*<br>*Cup & cone type fracture* | *Brittle fracture* |
| 135°C | *Cup & cone type fracture* | |
| 200°C | *Cup & cone type fracture* | |
| 300°C | *Cup & cone type fracture*<br>*Cup & cup type fracture* | |
| 350°C | *Cup & cup type fracture* | |
| 400°C | *Cup & cup type fracture* | |
| 600°C | *Chisel edge type fracture* | |
| 800°C | *Unidentified* | |

Table 3. Fracture types for the ring tensile tests of the Zircaloy-4 fuel cladding

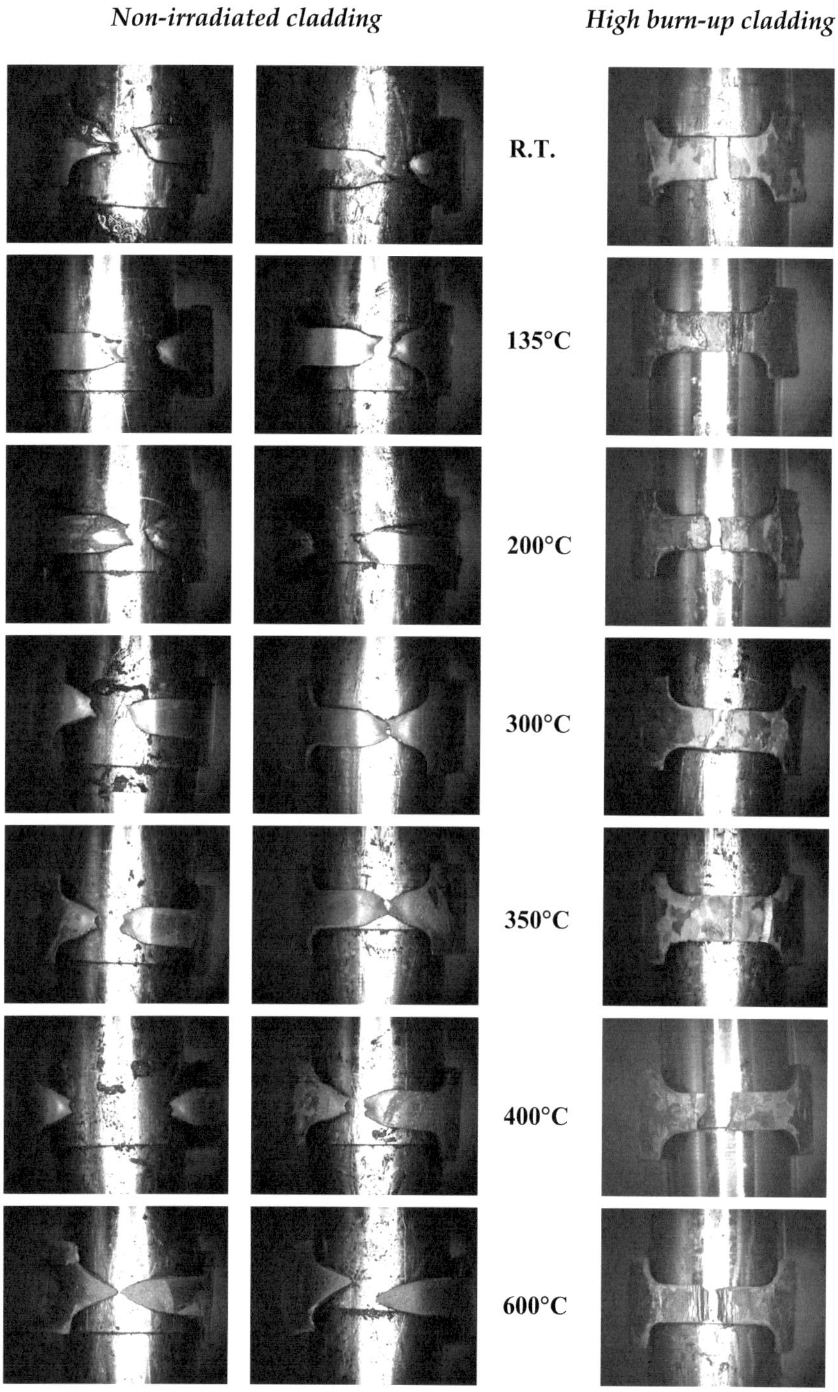


Fig. 8. Stereoscope photograph of the specimens after the ring tensile test

In conclusion, the fracture type of non-irradiated Zircaloy-4 cladding depends on the temperature, and the behavior is governed by a ductile fracture with a necking. On the contrary, the fracture patterns of the high burn-up Zircaloy-4 cladding showed completely different fracture patterns from the non-irradiated Zircaloy-4 cladding. The fracture type was observed to be vertical in the tensile direction without a necking at all of the test temperatures, which is convincing evidence of the brittle fracture behavior of high burn-up fuel cladding regardless of temperature.

This means that even at a high temperature, 600°C or 800°C, the fracture pattern showed a brittle fracture behavior. Accordingly, it was found that the high burn-up Zircaloy-4 cladding becomes very brittle even at the high temperatures achievable during a design-basis accident (Daum et al., 2002).

### 3.2 Evaluation of mechanical properties by high-speed burst test

Fig. 9 shows the time–pressure profile of the as-received Zircaloy-4 cladding under the rapid pressurization test. In the 350◦C test, the pressure increased prior to the test which is due to a thermal expansion of the hydraulic oil. It showed that the pressurization rate (slope at the initial part of the time–pressure curve) in this study was 5.4 GPa/s at room temperature and 3.1 GPa/s at 350◦C, where the pressurization rate at room temperature is 24,000 times higher than the conventional burst test. As the pressurization changes, the material properties also change. Ultimate hoop stress (UHS) at room temperature and 350◦C in this study were, respectively 1,067 and 620MPa, which are increased by 24.3 and 16.8% when compared to the conventional burst test. In the figure, theoretical burst behavior can be expected when the line was drawn from the elastic region and extended to the baseline, namely the 0MPa line. When we measure the interval between the intersection point and the point at a failure, the actual duration time of the fuel cladding during a rapid pressurization can be obtained. Duration time of the Zircaloy-4 cladding was 37.8 ms at room temperature and 32.5 ms at 350◦C, which closely simulates the power pulse of an actual RIA situation which is known as around 30–40 ms (MacDonald et al., 1970).

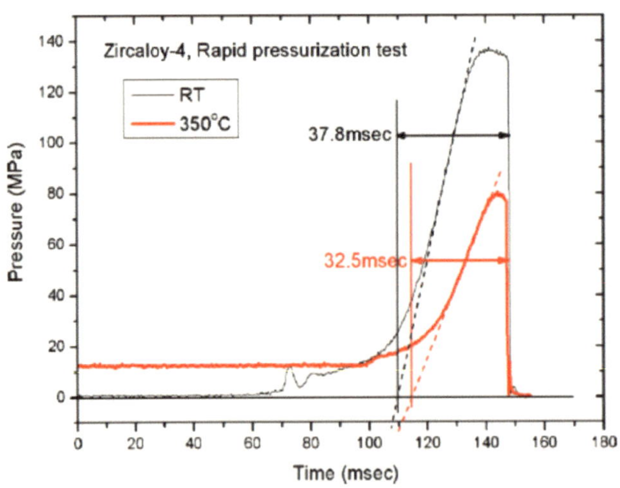


Fig. 9. Rapid burst properties of Zircaloy-4 cladding with the test temperature

Merging all burst tests into a single graph, the change in the material property as UHS with the pressurization is shown in Fig. 10. In the figure, there exists a linear relationship between the UHS and the pressurization rate. The slope between the UHS and pressurization rate was 0.03701 at room temperature and 0.02124 at 350◦C. From the result, a higher pressurization rate induces a higher strain rate that results in an increase of the UHS. Since the evaluation of the actual strain rate under biaxial burst is extremely difficult, it was tried to indirectly evaluate the strain rate of the cladding and compare to the other data (Kim et al., 2006).

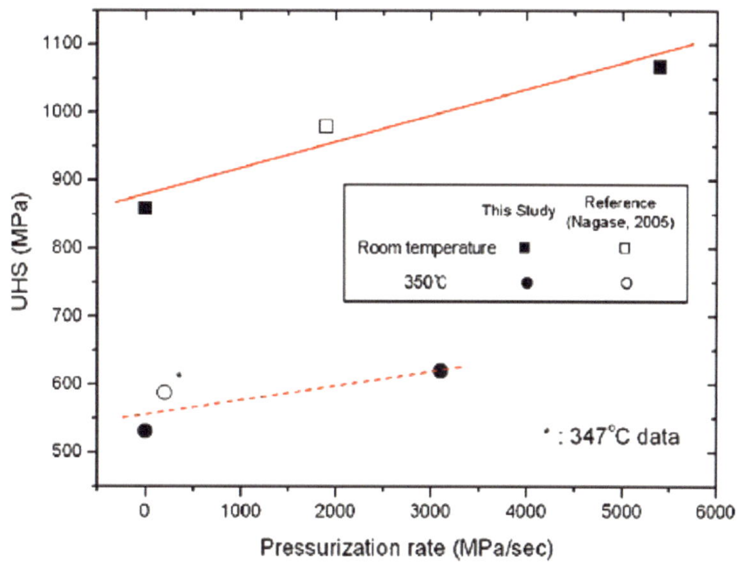


Fig. 10. Changes in the maximum hoop stress of Zircaloy-4 with the pressurization rate and the test temperature.

The result was shown in Fig. 11. As shown in the figure, increase of the maximum stress (both UTS in ring tensile test and UHS in burst test) with the strain rate is clearly shown. Such a strain rate hardening is known as the hindrance of dislocation movement caused by the dislocation multiplication. When an external strain is exerted, dislocations are continuously generated and diffused through the crystal structure to the grain boundary. If the rate of the dislocation generation exceeds the rate of diffusion to the grain boundary, the dislocation density increases to cause an increase in the strength (Adams, 1965). However, it is not clear why the maximum stress as well as the strain rate sensitivity of the burst test is higher than that of the tensile test. Further works related to the strain rate dependency and stress state of the zirconium cladding is needed.

During the waterside corrosion, the generated hydrogen is partly absorbed into the cladding. For hydrogen contents exceeding significantly the solubility limit, hydride precipitation is observed. These precipitated hydrides have a deleterious impact on the fuel cladding ductility (Kim et al., 2006). To investigate the effect of hydrogen, hydrogen was charged into the Zircaloy-4 cladding at a value of 300 and 600 ppm then a rapid pressurization test was carried out. Fig. 12 shows the room temperature rapid pressurization profile of the hydrogen-charged Zircaloy-4 cladding. UHS of the 300 ppm-charged specimen did not differ from the as-received condition except that the failure time was more or less shortened. In the case of the 600 ppm-charged specimen, it was so brittle that it failed at the elastic region. However, the effect of the hydrogen was soon eliminated when the test temperature was increased. Fig. 13 shows the UHS of the as-received and 600 ppm-charged specimens with the test temperature.

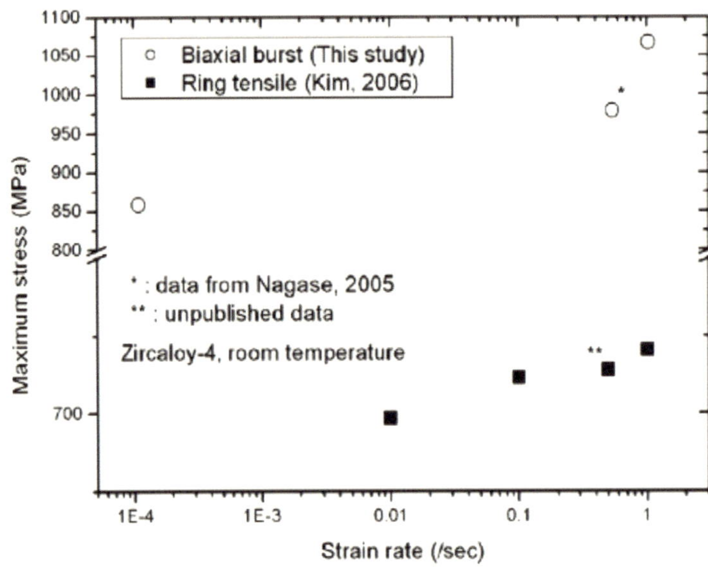


Fig. 11. Changes in the maximum stress of Zircaloy-4 with the strain rate

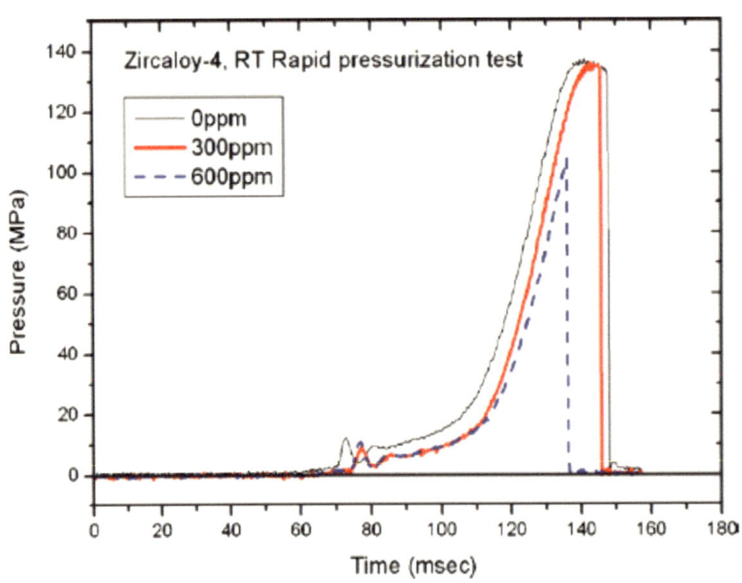


Fig. 12. Rapid pressurization properties of Zircaloy-4 cladding with the hydrogen content

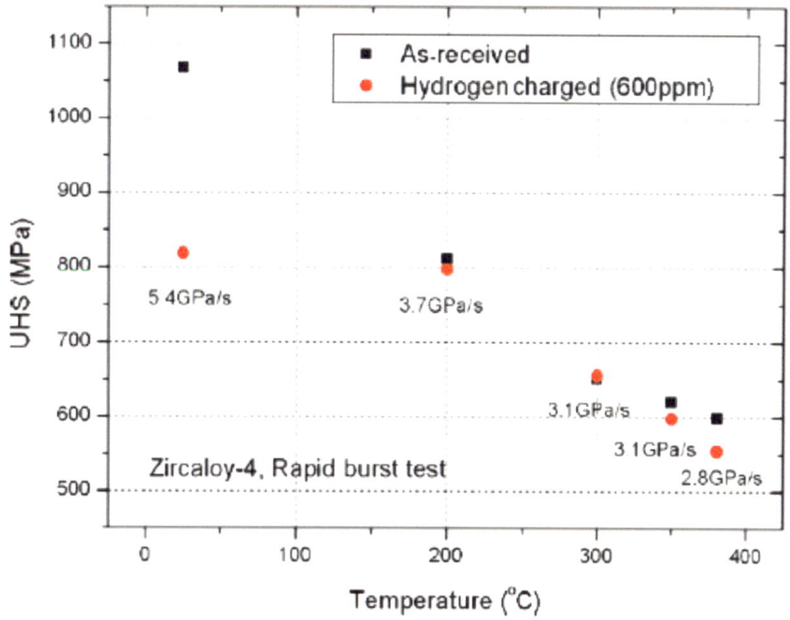


Fig. 13. Maximum hoop stress vs. test temperature

As the test temperature increases, the expansion of the cladding as well as the decrease in the viscosity of the hydraulic oil makes the pressurization rate become lower, which is regarded it as unavoidable. At room temperature, UHS of the as-received condition was 23.2% larger than that of the 600 ppm-charged specimen. When the temperature increases above 200 ◦C, their difference cannot be discriminated. At room temperature, the solubility limit of hydrogen is negligible and most of the hydrogen is located in the zirconium hydrides precipitates within the zirconium matrix. These hydrides are responsible for material embrittlement. When it strained at room temperature, separation between the hydride and the metal matrix as well as the failure of a brittle hydride itself will occur to reduce the strength of the zirconium cladding (Kim et al., 2006). When the test temperature increases, the property of the zirconium metal will be changed in that the ductility increases due to an increase in the test temperature. Thus it induces a similar mechanical behavior between the hydrogen-charged cladding and the as-received one. This is also implied from the fracture appearance as shown in the Fig. 14.

In the case of as-received cladding, whatever the testing temperature was, it showed a ductile rupture, creating a circumferentially ballooned state. Crack initiates and propagates along the axial direction above the maximum hoop stress, soon this sharp crack became blunt during a plastic deformation and it ceased to propagate. As-received specimen at 350◦C showed that the shape of the end part of the ballooned region was dull when compared to the room temperature and 200◦C condition, which implies that the ductility was increased to suppress the propagation of the axial crack at an elevated temperature. For the 600 ppm-hydrogen-charged cladding pressurized at room temperature, the cladding showed a brittle failure along the axial direction, indicating no plastic deformation. The crack propagates along the axial direction first like the as-received condition, however, this sharp crack did not become blunt at the cladding material and it rapidly propagated to induce a catastrophic, brittle failure. A void will nucleate preferentially at an interface between a hydride and a zirconium matrix to develop a crack.

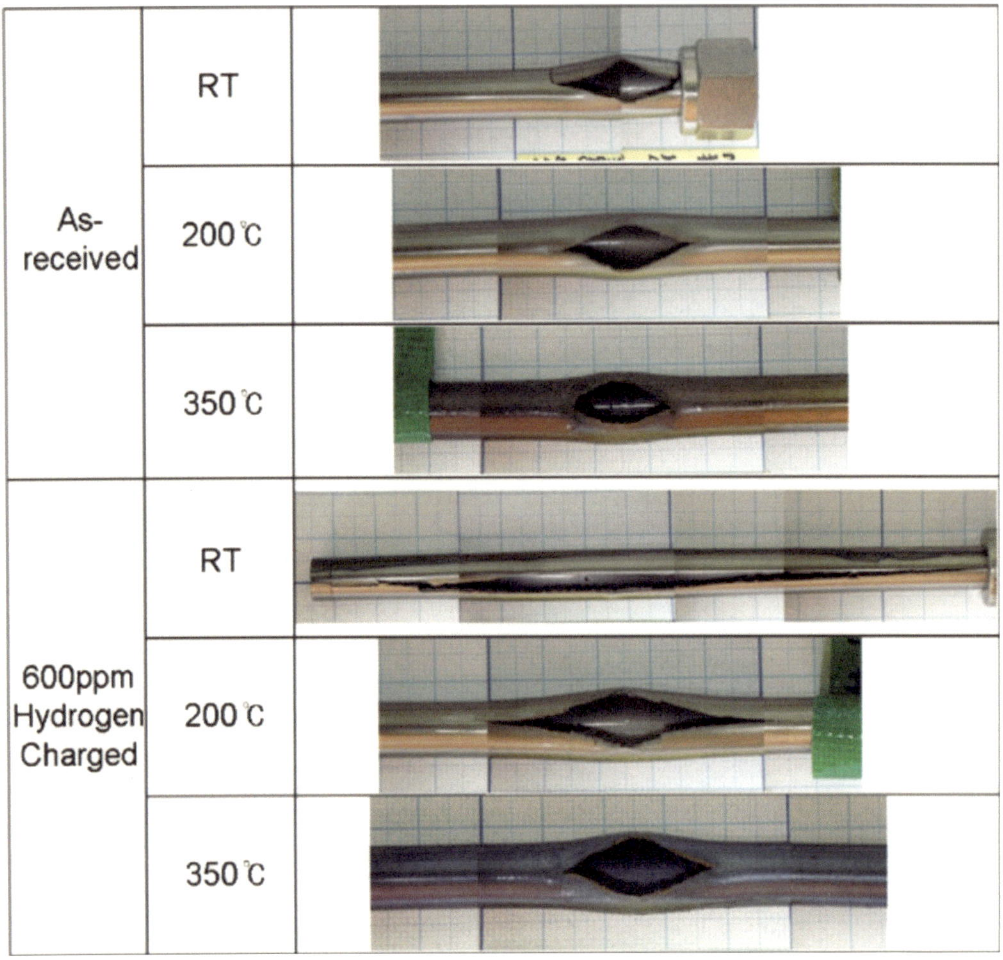


Fig. 14. Fracture appearance of Zircaloy-4 after the high-speed pressurization test

Void will continuously generate ahead of the crack tip when the crack propagates so that it leads to a catastrophic failure along the axial direction. In the case of the hydrogen-charged specimen burst at 200◦C, although the axial cracks developed at the side of the ballooned region, their size was not longer than that at the room temperature. In the case of the hydrogen charged specimen burst at 350◦C, the fracture appearance was no more different than that of the as-received one. In the room temperature, axial split was well developed at the end part cladding which indicates that residual metal is too brittle to suppress the axial crack propagation once the crack initiates at the metal–hydride interface. At the high temperature, such a catastrophic axial crack diminished. This seems that ductility of the metal matrix was so high that it can restrain the crack from propagating, leaving a fracture appearance similar to the as-received one.

## 4. Conclusion

Main outcomes from CABRI REP-Na program and the current status of CABRI international program (CIP) in water loop test program for investigating fuel and cladding behavior during RIA were presented.

On the basis of the ring tensile tests for the high burn-up Zircalay-4 cladding from Ulchin Unit 2 in Korea and the as-received non-irradiated Zircalay-4 cladding, the following conclusions were drawn. As a result, the mechanical properties are abruptly degraded beyond 600°C, which corresponds to a design basis accident condition such as a RIA. It was found that the un-irradiated fuel cladding showed ductile fracture behaviors such as 45° shear type fracture, cup and cone type fracture, cup and cup type fracture and chisel edge type fracture, while the high burn-up Zircalay-4 cladding showed a brittle fracture behavior even at the high temperatures (e.g. over 600°C) which are achievable during a RIA.

Rapid pressurization test was also carried out to evaluate the mechanical behavior of a cladding under a fast strain rate as well as a biaxial stress state to simulate an out-of-pile RIA behavior. The resuts shows that maximum hoop stress, pressurized at a rate of 5.4 GPa/s at room temperature and 3.1 GPa/s at 350 °C, increased by 24.3 and 16.8% when compared to the conventional burst test results. It was also revealed that failure mode switched from a ductile ballooning to a brittle failure which leads to an axial split of the cladding when the hydrogen was added at a nominal value of 600 ppm when tested at room temperature. As the test temperature increased, its effect was diminished.

## 5. Acknowledgment

This work was financially supported by the Ministry of Education, Science and Technology in Korea.

## 6. References

Adams, K. H. (1965). Dislocation Mobility and Density in Zinc Single Crystals. Ph.D. Thesis. California Institute of Technology

Averty X. et al. (2003). Tensile tests on ring specimens machined in M5 cladding irradiated 6 cycles, IRSN 2003/50

Chung, H. M. & Kassner, T. F. (1998). Cladding metallurgy and fracture behavior during reactivity-initiated accidents at high burnup. *Nuclear Engineering and Design*, Vol.186, pp. 411-427

Daum R. S. et al. (2002). On the Embrittlement of Zircaloy-4 Under RIA-Relevant Conditions. Zirconium in the Nuclear Industry : *Thirteenth International Symposium*, ASTM STP 1423, PA, pp.702-719

Glasstone, S. & Sesonske, A. (1991). *Nuclear Reactor Engineering*, 3rd ed., Krieger Publishing Company, Malabar, Florida, USA

Kim, et al. (2006). Effects of oxide and hydrogen on the circumferential mechanical properties of Zircaloy-4 cladding. *Nuclear Engineering and Design*, Vol.236, pp.1867–1873

MacDonald, P. E. et al. (1970). Assessment of light-water-reactor fuel damage during a reactivity-initiated accident. *Nuclear Safety*, Vol.21, No.5, pp.582–602

Meyer R. O. et al. (1986). A Regulatory Assessment of Test Data for Reactivity Initiated Accidents. *Nuclear Safety*, Vol.37, No.4, pp.271-288

Schmitz, F. & Papin, J. (1999). High burnup effects on fuel behavior under accident conditions: the tests CABRI REP Na, *Journal of Nuclear Materials*, Vol.270, pp. 55-64

# 8

# Introduction to Criticality Accident Evaluation

Yuichi Yamane
*Japan Atomic Energy Agency*
*Japan*

## 1. Introduction

For the use of nuclear energy, uranium must be processed to make a form to fit the purpose, such as a pellet. In nuclear fuel fabrication process, uranium is dissolved with nitric acid to make homogeneous solution. Uranium nitrate solution is full of water, which is the best moderator for neutron. That is a well-known reason of the criticality accident of nuclear fuel.

So far 22 criticality accidents have been reported (McLaughlin et al., 2000) including JCO accident in Japan. All cases took place with fuel solution or slurry except one case of metal fuel. It was not clear that what was actually happening during those criticality accidents except the JCO accident, for which a fission power profile was reproduced from a gamma-ray monitoring data (Tonoike et al., 2003). It is not easy to understand the phenomenon of criticality accident in detail, because it is a mixture of reactor physics, thermal dynamics and fluid dynamics.
In a criticality accident, the dose of the employee or public is the most important information. It is estimated from the amount of fission products produced during the criticality accident. The amount of fission products is proportional to the total number of fission, which is used to estimate the scale of the criticality accident as well.

Many kinetic methods have been developed for the estimation of the total number fission (Mather et al., 1984; Basoglu et al., 1998; Pain et al., 2001; Nakajima et al., 2002a, 2002b; Mitake et al., 2003). Some of them shows good agreement to experimental results (Miyoshi et al., 2009), however, Such methods requires rather high cost for the calculation and a lot of calculations are needed to find the response of the result to the variables such as temperature, input reactivity, etc.

Four simplified methods have been proposed to calculate the total fission number for a criticality accident (Tuck, 1974; Olsen et al., 1974; Barbry et al., 1987; Nomura & Okuno, 1995). Some are empirical equations and some are theoretical. Those simplified methods are easy to use, low cost and quick calculation, however, are known to overestimate the number of fission too much (Nakajima, 2003). Such overestimation could be reduced if we would focus to the detail power profile during the criticality accident.

The aim of the chapter is to introduce a concept of new method developed to evaluate the number of fission in a criticality accident, which is expected to give reasonable value, not too much overestimated, i.e. the estimated value is in the almost same order as the actual value.

The 1st section introduces the phenomena of the criticality accident with uranyl nitrate solution based on the TRACY experiment, which has been conducted by Japan Atomic Energy Agency (Nakajima et al., 2002c, 2002d, 2002e). The power profile is divided into three parts, the 1st peak, monotonically decreasing and plateau, for them the dominant mechanism and its time scale are different to each other. In the 2nd section, the condition characterizing a criticality accident is considered, such as temperature, reactivity temperature coefficient, water, cooling, etc. In the 3rd section, a new simplified method to evaluate the total fission number is described. The estimation is done for part-by-part by using equations differently introduced based on one-point kinetics equation or thermal equation (Yamane et al., 2007, 2008, 2009). In the 4th section, the new developing method is applied to some case to see its applicability.

## 2. Characteristics of criticality accident

In this section, the character of criticality accidents is explained. The most description is based on the data from TRACY experiments. At the first, the image of phenomenon, what happening, and its underlying physics are briefly introduced. Then, the conditions characterizing criticality accidents are described.

### 2.1 Phenomenon

A criticality accident occurs if enough amount of nuclear solution fuel such as uranyl nitrate solution is pumped into a tank with a shape not designed to avoid criticality. In most cases, very high energy caused by the fission of uranium is released in a moment, which is called "the first peak" of power profile. At the same time, the temperature of uranium and surrounding water is increased due to the released fission energy and the system becomes subcritical. After the first peak or at the same time, radiolytic gas void mainly due to the fast moving of fission product in the water is created and grows up. In any case, the system is approaching critical again, but if the system were disconnected thermally from surrounding materials, it would keep being subcritical. This phenomenon is a typical example of the system of uranyl nitrate aqueous solution. For the case of the largest total fission so far, criticality terminated after a large amount of water had been vaporized out. For the system of dilute plutonium solution, powder or metal, the phenomena may be different to each other.

### 2.2 Physics

Nuclear side of the phenomenon noted above is described by transport equation of neutron, which consists of neutron flux and its probability of reaction with nuclides. It can be applicable to any complicated condition but solving the equation for a complicated geometry, for example, requires a lot of computation power. Some assumptions, however, can reduce the complexity of neutron transport equation to make one-point kinetics equation (1). It has a simple form and even has the general solution for the simplest condition. It is enough for our purpose to understand the underlying physics of criticality accident.

$$\frac{dn}{dt} = \frac{\rho - \beta_{eff}}{\ell} n + \sum_{i=1}^{6} \lambda_i C_i$$
$$\frac{dC_i}{dt} = \frac{\beta_i}{\ell} n - \lambda_i C_i \quad (1)$$

where power, $n$, is used instead of the number of neutron, because they are proportional to each other and the equations are homogeneous. In the equation (1), $\rho = (k-1)/k$ is reactivity and $k$, neutron multiplication factor. $k=1$ means the number of neutron doesn't change in course of time. $k>1$ means its increase and $k<1$, decrease. They correspond to $\rho=0$, $\rho>0$ and $\rho<0$ each other. When a system is critical, its $\rho=0$ and $\rho<0$ stands for subcritical. Neutron multiplication factor, $k$, is a property of material and it is defined for infinite geometry. An actual system has a finite geometry and there must be some leak of neutrons from the system depending its shape. For such finite system, effective neutron multiplication factor, $k_{eff}$, is used instead of $k$. $C_i$ is the density of $i$-th delayed neutron precursor, which is a source of neutrons being released with a time constant $\lambda_i$, a decay constant of $i$-th delayed neutron precursor. The order of $\lambda_i$ is in the range of about 0.17s and 55s.

The ratio of the number of delayed neutrons to the total neutrons is denoted as $b_{eff}$, where suffix *eff* means that the leak of precursors and delayed neutrons from the fuel solution is considered. The system is delayed critical if $0<\rho< b_{eff}$; the system needs a help of delayed neutron to keep fission chain reaction and the fission power grows linearly. If $\rho> b_{eff}$, the system is prompt critical; fission chain reaction continues only with prompt neutrons and the power increases exponentially. $\rho$ divided by $b_{eff}$ is used for convenience and comes with unit "$"; i.e. more than 1$ excess reactivity corresponds to prompt critical. $\ell$ is prompt neutron life time. The time scale of neutronics is very small because $\ell$ is small. For example, $\ell$ is about $5 \times 10^{-5}$s for TRACY.

During a criticality accident, most of released fission energy is consumed to increase the temperature of uranium and other materials. Some equations such as thermal conduction equation, heat diffusion equation or Fourier's law can describe the distribution of thermal energy. The temperature of uranium is important because it has reactivity effect well-known as "Doppler effect," which can change the system's total reactivity, mainly decreases it. The thermal expansion of the solution has the same effect as well.

After a fission, high energy fission products run into water, excite or break water molecules and the overlapping of those exciting atoms and ions make voids in the fuel solution. Such radiolytic gas void usually has a negative reactivity effect; a lot of void can make the system subcritical. Boiling void has the same reactivity effect. The motion of void is a matter of fluid dynamics and can be described as multi phase flow. The knowledge of the elementary step of creating void is not enough, however, some models are used tentatively (Nakajima et al., 2002).

### 2.3 Power

For criticality accident, in other words, transient criticality, power profile has its unique pattern depending on the condition of the system. For an example, a typical power profile is explained based on TRACY experiment.

Figure 1 shows a power profile data obtained from a transient criticality experiment with 1.5$ excess reactivity inserted by pumping fuel solution into the TRACY core tank. At the first, the power increases exponentially and suddenly decreases to make a peak in the power profile. Then, the power decreases monotonically. Finally it recovers and keeps some value as plateau shown in Fig.1.

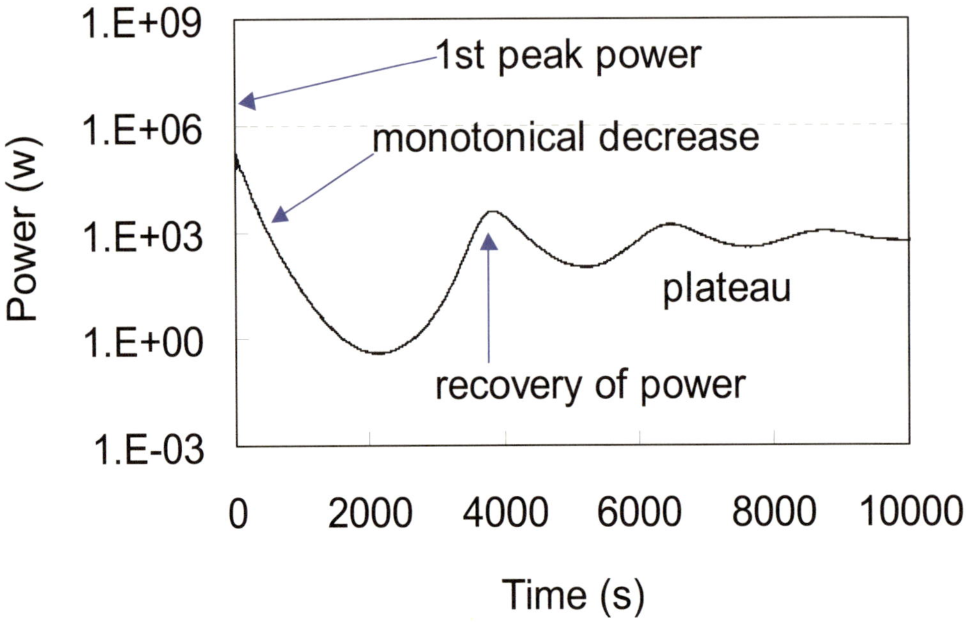


Fig. 1. Example of power profile observed by using TRACY for the case of 1.5$ inserted by feed of nuclear fuel solution.

Around the first peak for the case more than 1$ excess reactivity inserted instantaneously, the power profile is described as follows;

$$\frac{dn}{dt} = \frac{\rho - \beta_{eff}}{\ell} n \qquad (2)$$

Because the width of the 1st peak is less than 0.1s, which is less than the smallest $l_i$, the change in C can be ignored.

Monotonic decrease of the power is driven by the decay of neutron precursors and described as follows;

$$n = n_{pe} \exp(-\lambda_1 t) \qquad (3)$$

where $\lambda_1$ is the smallest decay constant.

The power at plateau corresponds to the cooling rate of the fuel solution. For TRACY, during several hours from the first, the thermal conduction from the fuel solution to the stainless steel tank and its support plays a dominant role.

It should be mentioned that another pattern is observed for another condition such as initially boiling, dilute plutonium solution, etc.

### 2.4 Energy-fission

Released fission energy is consumed to increase the temperature of uranium and other nuclides except some loss. one fission gives rise to about 200MeV energy and $10^{18}$ fission is almost equal to 32MJ. The number of fission at the largest criticality accident so far is $4x10^{19}$. That is, however, a rare case. For the second largest and other cases, the number of fission is $3x10^{18}$ or less.

The number of fission is used to evaluate the public dose around a nuclear facility. For example, for the design of Rokkasyo fuel reprocessing facility, $10^{19}$ fission is used for DBE, Design Basis Event, which is a postulated event to evaluate the safety of the design of the facility, $10^{20}$ for SEA, Siting Evaluation Accident, which is an postulated accident to evaluate the safety of the public around the facility (Working Group on Nuclear Criticality Safety Data, 1999).

### 2.5 Temperature

The temperature of nuclear fuel solution increases when a criticality accident occurs. Figure 2 shows a typical example of the temperature profile measured at TRACY. The difference value from the initial temperature is determined the released fission energy and the specific

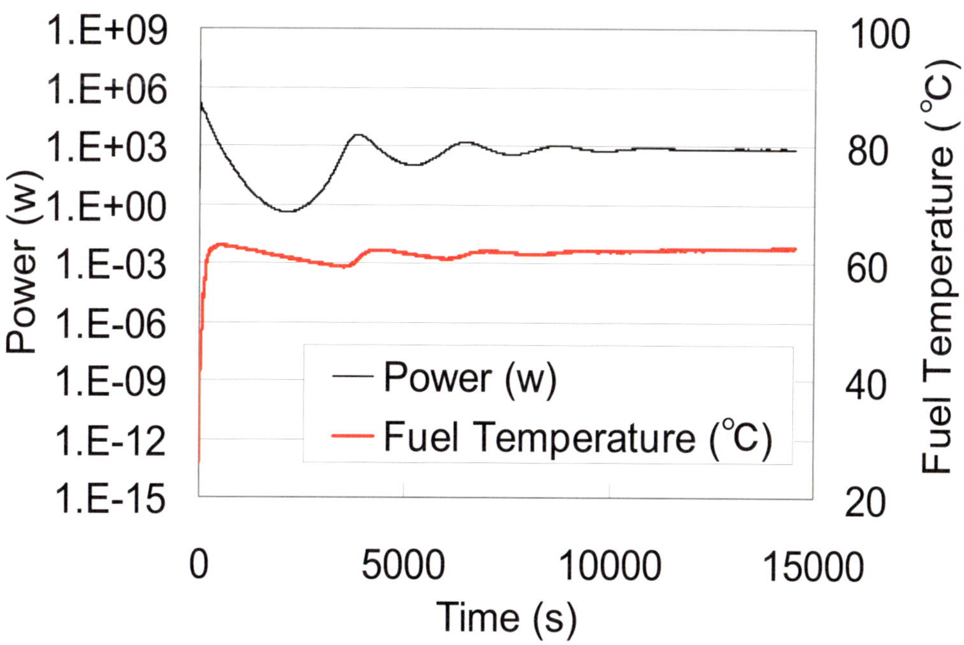


Fig. 2. Example of fuel temperature profile with power profile for the case of 1.5$ inserted by feed of nuclear solution fuel.

heat of the fuel solution and it gives rise to reactivity feedback. The ratio of the feedback reactivity to the temperature difference is called "reactivity temperature coefficient." More precisely, feedback reactivity is a function of temperature difference and described as follows;

$$\rho = \sum_i \alpha_i \Delta T^i \tag{4}$$

where $\alpha_i$ denotes i-th order reactivity temperature coefficient. For uranyl nitrate solution, the temperature feedback reactivity has negative value and is a non-linear function of temperature difference usually.

### 2.6 Pressure

In the fuel solution, rapid increase of temperature gives rise to rapid expansion of the solution. That is observed as pressure. Some pressure can be observed and measured at pulse withdrawal mode experiment of TRACY, for that the neutron absorber rod in the centre of the core tank is instantaneously withdrawn. Pressure can observed clearly at slow transient experiment as well.

## 3. Simplified evaluation of fission yield

In this section, a new simplified evaluation method for the number of fission released at a criticality accident is proposed, which is consist of three parts.

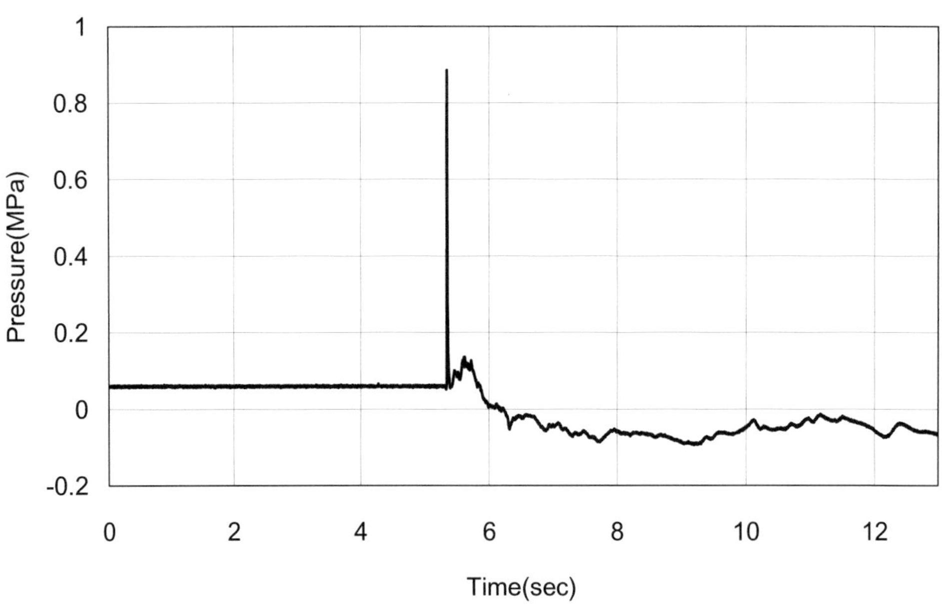


Fig. 3. Example of pressure profile measured at TRACY experiment for the case of 2.9$ inserted instantaneously by pulsed withdrawal of the transient rod (Nakajima et al., 2002a).

### 3.1 Around first peak power

If all of the excess reactivity is given to the system in pseudo-steady manner, a power peak like the one shown in section 2.3 cannot be observed. In most actual cases, the excess reactivity was inserted in a rate and that caused an exponential growth of power. It is known that if the excess reactivity is given instantaneously the value of the 1st peak power, $n_p$, and the energy released in the 1st peak, $E_p$, are denoted as follows;

$$n_p = \frac{(\rho - \beta)^2}{2\alpha K \ell} \tag{5}$$

$$E_p = \frac{2(\rho - \beta)}{\alpha K} \tag{6}$$

where $\alpha$ is reactivity temperature coefficient, $K$ reciprocal heat capacity. These expressions were introduced analytically by Nordheim and Fuchs (Fuchs, 1946; Nordheim, 1946). The estimated value for the released energy during the 1st power peak using equation (6) was compared to the experimental data of TRACY experiment and it was found that they showed good agreement to each other for the excess reactivity greater than 2$ as shown in Fig. 4.

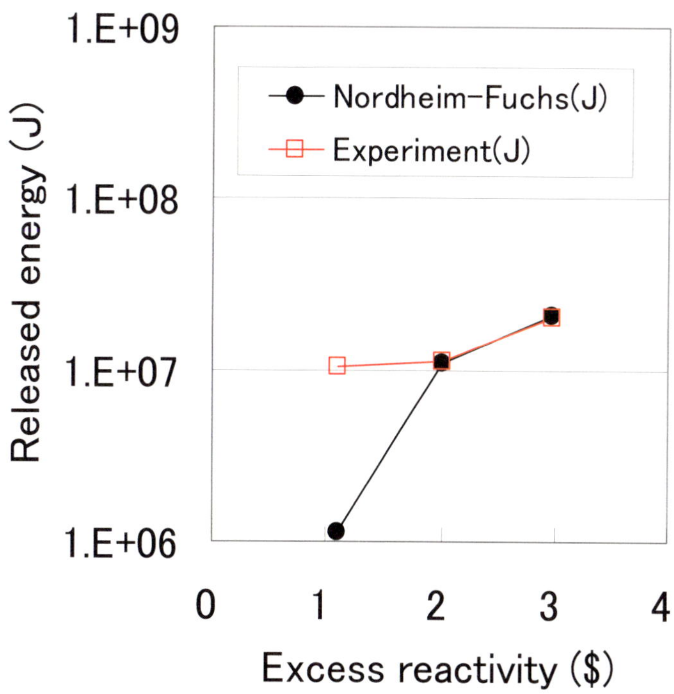


Fig. 4. Estimated value based on Nordheim-Fuchs model compared to TRACY experimental data for released energy during the 1st power peak (Yamane, 2009).

Because the instantaneous reactivity insertion corresponds to the largest insertion rate, the expression (6) gives the largest energy for the same excess reactivity case. And for our purpose, to evaluate the number of fission in the simplest manner, it is enough useful.

### 3.2 Monotonically decreasing

After the 1st peak, the power decreases monotonically. Based on one-point kinetics, the simple expression for the power $n$ during the monotonically decreasing can be introduced.

Let us start with the following equations (Yamane, 2009);

$$\frac{dn}{dt} = \frac{\rho - \beta_{eff}}{\ell} n + \lambda C$$
$$\frac{dC}{dt} = \frac{\beta_{eff}}{\ell} n - \lambda C \tag{7}$$

where $\lambda$ is the average of $\lambda_i$ in equation (1);

$$\lambda = \left[ \frac{1}{\beta} \sum_{i=1}^{6} \frac{\beta_i}{\lambda_i} \right]^{-1} \tag{8}$$

For the excess reactivity larger than 1$, integrating the equations (7) gives

$$n = n_{pe} \exp\left( \frac{\lambda}{\beta / \rho - 1} t \right) \tag{9}$$

for the power after the 1st peak, where $n_{pe}$ denotes the power at the end of the 1st peak. For the uranyl nitrate solution used for TRACY experiment, the value of $\lambda$ is 0.08 (1/s), $\beta$, 0.0076. During monotonically decreasing of the power, the total reactivity is negative and its nominal value is large, for example, minus several ten $. So, the equation (8) can be denoted in simpler form as follows;

$$n = n_{pe} \exp(-\lambda t) \tag{10}$$

Monotonically decreasing of the power continues for long time, hundreds or more, and in such time scale, the smallest decay constant $\lambda_1$ is dominant. So, released energy during monotonically decreasing, $E_d$, is denoted as follows;

$$E_d = n_{pe} / \lambda_1 \left(1 - \exp(-\lambda_1 t_2)\right) \tag{11}$$

where $t_2$ is the time for monotonically decreasing of the power, $\lambda_1$, the smallest decay constant, 0.0127s for TRACY condition. Because the contribution of the power to the released energy is negligibly small for large $t_2$, $t_2$ can be $\infty$ and we have;

$$E_d = \frac{n_{pe}}{\lambda_1} \tag{12}$$

The power profile calculated by using the equation (10) is compared to TRACY experimental data as shown in Fig. 5. The calculation reproduces experimental power profile well. The difference after 400 seconds is due to cooling effect. For the released energy, as shown in Fig. 5, the value estimated by using the equation (12) shows good agreement to the experimental value.

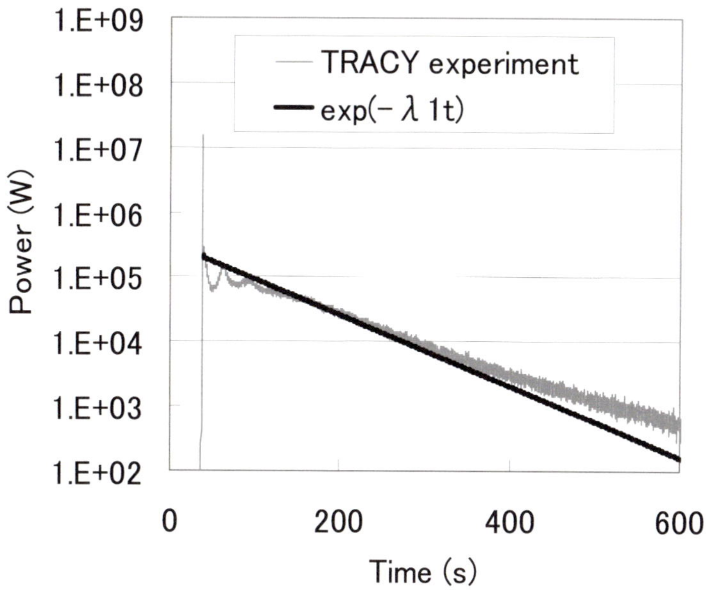


Fig. 5. Estimated power profile based on equation (10) compared to TRACY experimental data during monotonically decreasing of the power (Yamane, 2009).

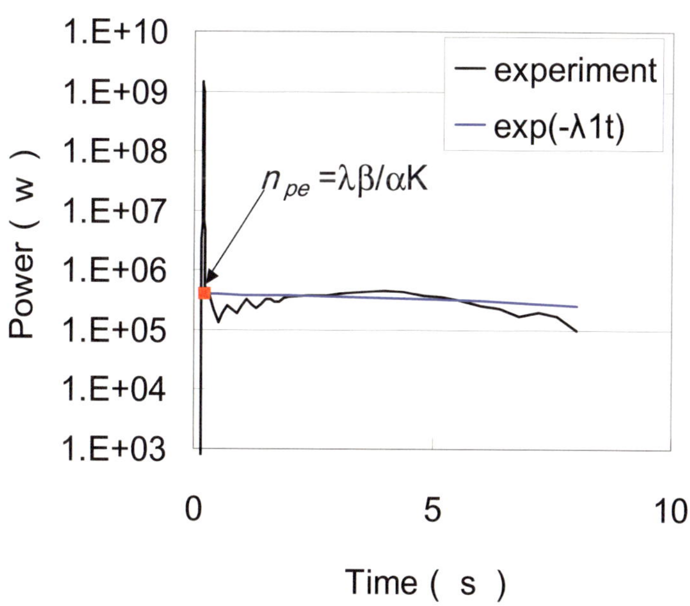


Fig. 6. Good case for estimated power profile based on equations (10) and (13) compared to TRACY experimental data during monotonically decreasing of the power (Yamane, 2009).

A candidate of the theoretical expression of $n_{pe}$ has been introduced as follows;

$$n_{pe} \approx \frac{\lambda\beta}{\alpha K} \tag{13}$$

The calculated value using the equation (13) shows good agreement to the experimental value for some cases as shown in Fig.6, however, that is not always. As shown in Fig.7, there is some clear difference between those values. The expression of $n_{pe}$ can be much improved.

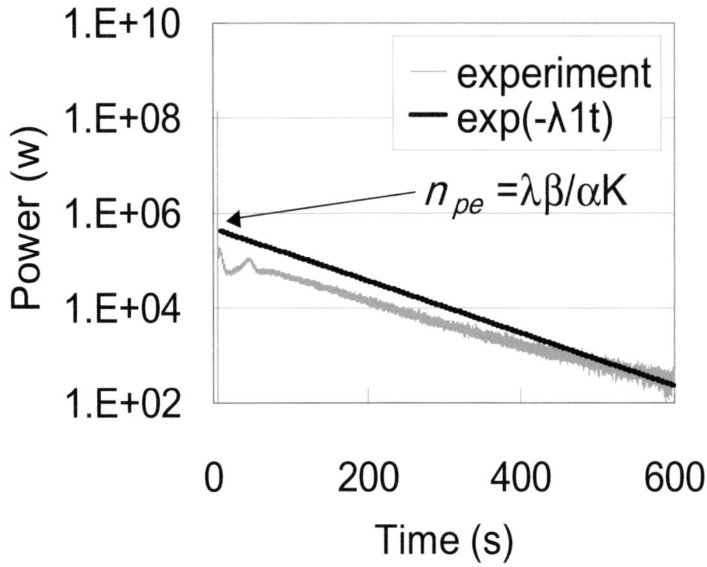


Fig. 7. Not good case for estimated power profile based on equations (10) and (13) compared to TRACY experimental data during monotonically decreasing of the power (Yamane, 2009).

### 3.3 Recovery and plateau

When the power decreases and it is very low, the fuel solution becomes being cooled and as its temperature decreasing the system becomes criticality again. Such cooling effect is mainly due to thermal conduction between the fuel solution and the container or surrounding materials. The fission power reaches such cooling power in course of time. As for TRACY, such cooling power is about 1kW as seen in Fig.6. If the cooling power is calculated by using CFD code in advance, the fission energy being released can be estimated easily.

It has been confirmed by calculation that the cooling effect is mainly due to thermal conduction between the fuel solution and the container. A kinetics calculation by using AGNES code into which a thermal conduction model based on Fourier's law has been implemented is compared to an experiment by using TRACY as shown in Fig.8. And it can be seen that the calculation reproduced the power profile obtained experimentally very well.

If the tank which contains the fuel solution is cooled forcibly, the fission power is the same as its cooling power. For example, the tank into which the fuel solution was poured had a water jacket in the JCO accident. The fission power kept high value because the cooling system kept working. After cutting the pipe between the jacket and the water supplier, the fission power got down.

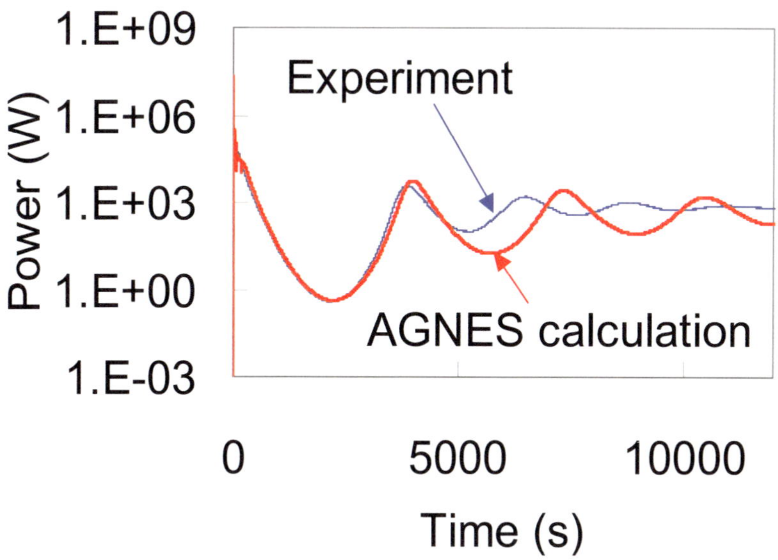


Fig. 8. Not good case for estimated power profile based on equations (10) and (13) compared to TRACY experimental data during monotonically decreasing of the power (Yamane, 2008).

### 3.4 New method

The total number of fission can be estimated as the sum of three values described above. Table 1 shows the whole expressions. To use these expressions, the excess reactivity, $\rho$, must be known in advance. In the case of planning the nuclear facility, it is common that credible maximum excess reactivity is calculated by using a montecarlo code or a deterministic code. In the case of criticality accident, such calculation should be done to confirm the effect of counter action or to understand what is happening. Anyway, it can be expected that we can obtain an estimated excess reactivity.

| Range | Expression |
|---|---|
| Around 1st peak | $E_p = \dfrac{2(\rho_\$ - 1)}{\alpha K}$ |
| Monotonically decreasing | $E_d = \dfrac{n_{pe}}{\lambda_1} \quad n_{pe} \approx \dfrac{\lambda\beta}{\alpha K}$ |
| Plateau | $E_c$ = Cooling power x duration time |
| Total | $E_p+E_d+E_c$ |

Table 1. New expression for simplified estimation of fission yield. $\rho_\$$ is excess reactivity in the unit $.

## 4. Example of application

The method proposed in the previous section was applied to a TRACY experiment to evaluate it. Here considered are two cases, fast and slow transient cases. Before going to evaluation, the outline of TRACY experiment is explained.

### 4.1 TRACY experiment

TRACY, transient experiment critical facility, is a critical assembly which fuel is uranyl nitrate solution (Nakajima et al., 2002a, 2002b, 2002c). Its enrichment of $^{235}U$ is 98.9% and uranium concentration is in the range of 375 to 433 g/Lit. Its free nitric acid molarity is in the range of 0.6 to 0.9 mol/Lit. Such solution is contained in a cylindrical tank of SUS and its inner diameter is 52cm as shown in Fig. 9. A guide tube for a neutron absorber rod, "transient rod," is in its centre and its outer diameter is 7.6cm. For an transient experiment, reactivity is given by three ways such as (1)pulsed withdrawal of the transient rod, Pulse Withdrawal mode, (2)slow withdrawal of the transient rod, Ramp Withdrawal mode, (3)pumping the fuel solution from the bottom of the tank, Ramp Feed mode.

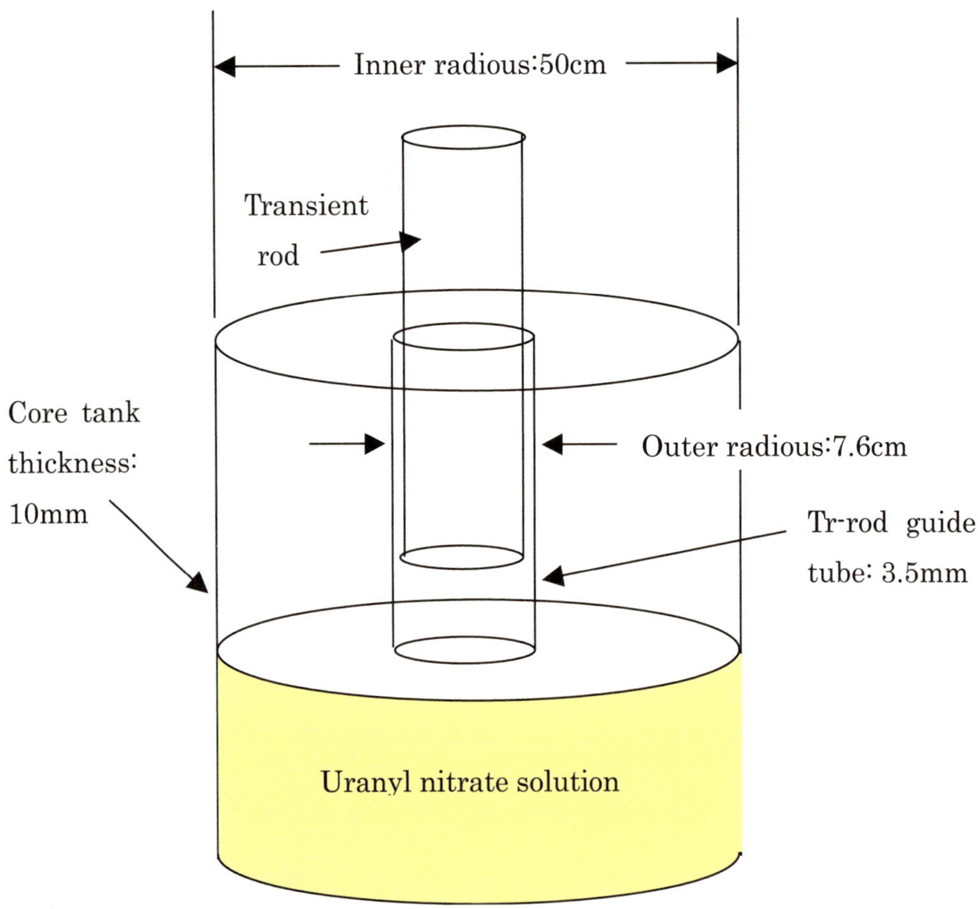


Fig. 9. Schematic view of TRACY

### 4.2 Fast transient case

For TRACY experiment R203, 2.97$ was inserted by pulsed withdrawal of the transient rod. The estimated values to use the simplest method are plotted in Table 2.

| Parameter | Estimated value |
|---|---|
| $\rho$ | 2.97$ |
| $\beta_{eff}$ | $7.6 x 10^{-3}$ |
| $-\alpha$ | $-6.3 x 10^{-2}$ $/°C |
| K | $2.0 x 10^{-6}$ °C/J |
| $\lambda$ | $8.0 x 10^{-2}$ 1/s |
| $\lambda_1$ | $1.27 x 10^{-3}$ 1/s |

Table 2. Parameters used to apply new method to R203.

For R203, free excursion was terminated 8 seconds after the insertion of reactivity as shown in Fig. 10. To consider the effect of the termination in short time, the equation (11) should be used instead of the equation (12). For TRACY experiment, reactivity temperature coefficient, $\alpha$, should be multiplied by 1.5 to consider the effect of the temperature distribution in the solution at the 1st peak of power. For this case, $E_c$ is zero, because the effect of cooling didn't appear. The result of calculation and the experiment are shown in Table 3. For the total number of fission, new method provides almost same value as the experiment. If we use the equation (12), $E_d = 1.2 x 10^{17}$ and the total number of fission is estimated to be $7.7 x 10^{17}$. That is enough close to the experimental value for our purpose.

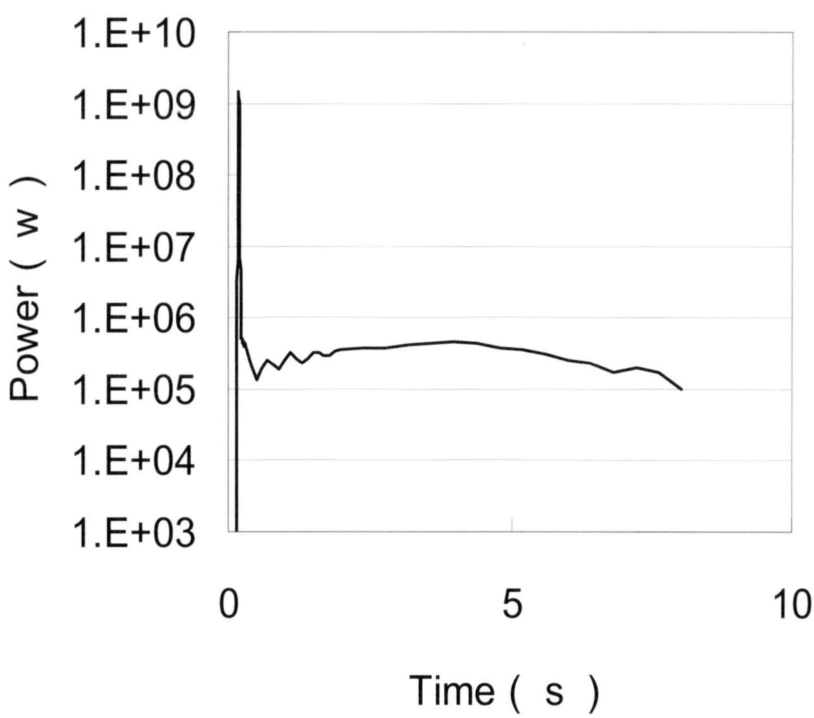


Fig. 10. Power profile of R203 (Nakajima et al., 2002a).

| | New method | Experiment |
|---|---|---|
| Total number of fission | $6.5 \times 10^{17}$ | $6.3 \times 10^{17}$ |
| $E_p$ | $6.5 \times 10^{17}$ | - |
| $E_d$ | $1.2 \times 10^{15}$ | - |
| $E_c$ | 0 | - |

Table 3. Simplified method applied to fast transient case of TRACY experiment R203.

### 4.3 Slow transient case

For TRACY experiment R164, 1.52$ was inserted by ramp feed of fuel solution. The estimated values to use the simplest method are plotted in Table 4.

| Parameter | Estimated value |
|---|---|
| $\rho$ | 1.52$ |
| $\beta_{eff}$ | $7.6 \times 10^{-3}$ |
| $-\alpha$ | $-6.3 \times 10^{-2}$ $/°C |
| K | $1.8 \times 10^{-6}$ °C/J |
| $\lambda$ | $8.0 \times 10^{-2}$ 1/s |
| $\lambda_1$ | $1.27 \times 10^{-3}$ 1/s |

Table 4. Parameters used to apply new method to R164.

For R164, free excursion continued about 15000s as shown in Fig. 11. In Fig. 11, the power decreases monotonically until about 2500s and its recovery and plateau can be seen between 4000s and 15000s. To estimate $E_p$, reactivity temperature coefficient, $\alpha$, should be multiplied by 1.5 to consider the effect of the temperature distribution in the solution at the 1st peak of power for TRACY. For this case, $E_c$ is about 1kw as seen in Fig. 11. The result of calculation and the experiment are shown in Table 5. For the total number of fission and $E_p$+$E_d$, new method provides a value close to the experiment. That is enough close to the experimental value for our purpose.

| | New method | Experiment |
|---|---|---|
| Total number of fission | $6.0 \times 10^{17}$ | $6.9 \times 10^{17}$ * |
| $E_p + E_d$ | $2.6 \times 10^{17}$ | $5.3 \times 10^{17}$ * |
| $E_p$ | $1.9 \times 10^{17}$ | - |
| $E_d$ | $6.9 \times 10^{16}$ | - |
| $E_c$ | $3.4 \times 10^{17}$ | - |

*tentative value

Table 5. Simplified method applied to fast transient case of TRACY experiment R164.

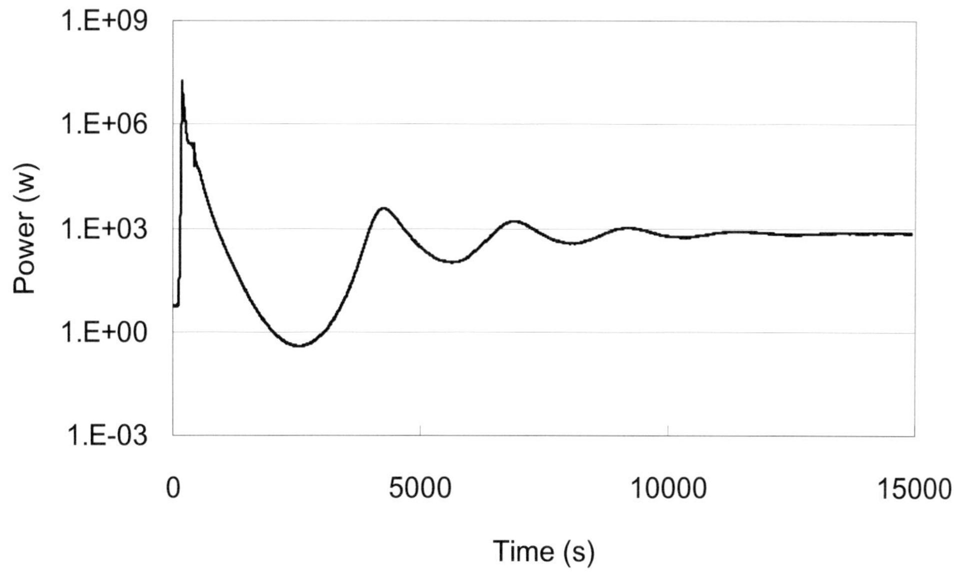


Fig. 11. Power profile of R164 (Nakajima et al., 2002c).

## 5. Conclusion

A concept for new simplified method to evaluate fission yield at a criticality accident was introduced and explained briefly. Two examples of its application were compared to experiments to show its applicability. The results showed the estimated values were enough close to the experimental values. There still are some points to be developed or improved; for example, $n_{pe}$ can be much accurate, the power during plateau should be denoted in a simple form, etc.

## 6. Acknowledgment

The author greatly appreciate Dr. Uchiyama who prompted to write this chapter, Prof. Nakajima who is giving useful suggestion always, Mr. Futagami who is a great help to measure and record the experimental data, and staffs committed to the operation of TRACY for their dedicated efforts.

## 7. References

Barbry F. et al. (1987). *Criticality Accident Studies in France: Experimental Programs and Modelisation*, Proc. Int. Seminar on Nuclear Criticality Safety, ISCS'87, Tokyo, Japan, October 19-23, 1987, p. 423.

Basoglu B., et al., (1998). *Development of a New Simulation Code for Evaluation of Criticality Transients Involving Fissile Solution Boiling*, JAERI-Data/Code 98-011.

Mather D.J., et al. (1984). *CRITEX – A Computer Program to Calculate Criticality Excursions in Fissile Liquid Systems*, SRD R 380.

McLaughlin T.P. et al. (2000). *A Review of Criticality Accidents 2000 Revision,* LA-13638.

Mitake S., et al. (2003). *Development of INCTAC code for Analyzing Criticality Accident Phenomena,* Proc. 7th International Conference on Nuclear Criticality Safety (ICNC2003), JAERI-Conf 2003-019.

Miyoshi Y., et al. (2009). *Inter-code Comparison Exercise for Criticality Excursion Analysis,* OECD/NEA No.6285.

Nakajima K., et al. (2002a) *A Kinetics Code for Criticality Accident Analysis of Fissile Solution Systems: AGNES2,"* JAERI-Data/Code 2002-004.

Nakajima K., T. Yamamoto and Y. Miyoshi, (2002b). *Modified Quasi-Steady-State Method to Evaluate the Mean Power Profiles of Nuclear Excursions in Fissile Solution,* J. Nucle. Sci. Technol., 39[11], 1162.

Nakajima K., et al. (2002c). *TRACY Transient Experiment Databook 1)Pulse Withdrawal Experiment,* JAERI-Data/Code 2002-005.

Nakajima K., et al. (2002d). *TRACY Transient Experiment Databook 2)Ramp Withdrawal Experiment,* JAERI-Data/Code 2002-006.

Nakajima K., et al. (2002e). *TRACY Transient Experiment Databook 3)Ramp Feed Experiment,* JAERI-Data/Code 2002-007.

Nakajima K. (2003). *Applicability of Simplified Methods to Evaluate Consequences of Criticality Accident Using Past Acident Data,"* Proc. of 7th International Conference on Nuclear Criticality Safety, ICNC2003.

Nomura Y. and Okuno H., (1995). *Simplified Evaluation Models for Total Fission Number in a Criticality Accident,* Nucl. Technol., 109, 142.

Olsen A.R. et al. (1974). *Empirical Model to Estimate Energy Release from Accidental Criticality,* Trans. Am. Nucl. Soc., 19, 189.

Pain C.C., et al. (2001). *Transient Criticality in Fissile Solutions Compressibility Effects,* Nuc. Sci. and Eng., Vol.138 pp.78-95.

Tonoike K. et al. (2003). *Power Profile Evaluation of the JCO Precipitation Vessel Based on the Record of the Gamma-ray Monitor,* Nucl. Technol. Vol.143, pp.364-372.

Tuck G., (1974). *Simplified Methods of Estimating the Results of Accidental Solution Excursions,* Nucl. Technol., 23, 177.

Yamane Y., et al. (2003). *Transient characteristics observed in TRACY supercritical experiments,* Proceedings of 7th International Conference on Nuclear Criticality Safety (ICNC2003).

Yamane Y., (2007). *Approximate Expression for Maximum Power at Slow Transient Criticality Accident,* Proc. 2007 Fall Meeting of AESJ.

Yamane Y., (2008). *Long-time Power Profile until Re-criticality and Recovery of Power in Criticality Accident,* Proc. 2008 Fall Meeting of AESJ.

Yamane Y., (2009). *Power Profile and Released Energy after First Peak in Criticality Accident,* Proc. 2009 Fall Meeting of AESJ.

Working Group on Nuclear Criticality Safety Data, (1999). *Nuclear Criticality Safety Handbook Version 2,* JAERI-1340.